COMPUTER COMMUNICATION
SYSTEMS
VOLUME 2

COMPUTER COMMUNICATION SYSTEMS

VOLUME 2

Principles
Design
Protocols

Henri Nussbaumer

EPF, Lausanne

Translated by
John C.C. Nelson
University of Leeds

JOHN WILEY & SONS
Chichester · New York · Brisbane · Toronto · Singapore

Other Wiley Editorial Offices

John Wiley & Sons, Inc., 605 Third Avenue,
New York, NY 10158-0012, USA

Jacaranda Wiley Ltd, G.P.O. Box 859, Brisbane,
Queensland 4001, Australia

John Wiley & Sons (Canada) Ltd, 22 Worcester Road,
Rexdale, Ontario M9W 1L1, Canada

John Wiley & Sons (SEA) Pte Ltd, 37 Jalan Pemimpin 05-04,
Block B, Union Industrial Building, Singapore 2057

Originally published under the title
'Teleinformatique' volume 2, by
Presses polytechniques romandes

Library of Congress Cataloging-in-Publication Data:

Nussbaumer, Henri J.
    [Téléinformatique, English]
    Computer communication systems / Henri Nussbaumer ; translated by
John C. C. Nelson.
        p.   cm.
    Translation of: Téléinformatique.
    Includes biographical references
    Contents: v. 1. Data circuits, error detection, data links — v.
2. Principles, design, protocols.
    ISBN 0 471 92379 6 (v. 1) — ISBN 0 471 92495 4
(v. 2)
    1. Computer networks.   I. Title.
TK5105.5.N8713 1990
004.6—dc20   89-24871
                                                      CIP

British Library Cataloguing in Publication Data:

Nussbaumer, Henri
    Computer communication systems.
    Vol. 2, Principles, design, protocols
    1. Computer systems. Networks design
    I. Title
    004.6′5

    ISBN 0 471 92495 4

Typeset by Photo·graphics, Honiton, Devon
Printed in Great Britain by Courier International, Tiptree, Essex

CONTENTS

INTRODUCTION

OBJECTIVES OF THE WORK

This work is the second volume of a series which gathers together the main aspects of teleprocessing in a single presentation. It is derived from courses given to computer engineering students at the Swiss Federal Institute of Technology, Lausanne, and is intended for those who have a basic knowledge of data processing and who wish to deepen their knowledge in the area of teleprocessing. This series presents the principles of teleprocessing, the techniques used for designing and modelling networks, and the communications protocols.

This volume is published jointly with the first which presents an introduction to teleprocessing, data circuits, error detection and correction and line procedures. These first two volumes treat the problems of modelling and describing the low-level protocols which extend from the transmission of bits on lines to end-to-end transport between terminal equipment. Subsequent volumes will be devoted to local area networks and high-level protocols.

GENERAL ORGANIZATION OF THE VOLUME

Chapter 1 is devoted to the analysis and design of teleprocessing systems. Our aim is to provide the reader with the basic elements which are necessary to undertake system modelling and are generally applicable to data processing systems. This has led to treatment of various topics such as queues, analysis of concentrators and optimization of network topology.

Chapter 2 presents routing and congestion control techniques in packet switching networks. It concludes with a description of the X.25 network protocol together with a brief introduction to the ISO connectionless network protocol.

Chapter 3 treats the transport protocols which are operated to ensure reliable transfer of information through a network. It gives the general principles of end-to-end transport between terminal equipment and presents the various aspects of the ISO transport protocol.

CONVENTIONS

The work consists of volumes represented by Roman numbers (e.g. Vol. I). Each volume is divided into chapters represented by an Arabic number (e.g. Chapter

3). Each chapter is divided into sections represented by two Arabic numbers separated by a point (e.g. Section 3.4) and each section is subdivided into sub-sections represented by three Arabic numbers separated by two points (§3.4.2). In the case of reference to a part of the current volume, its number is omitted.

A term appears in bold italic on its first occurrence in the text. Important terms or concepts are emphasized in italic. Terms in italic appear in the alphabetical index.

Equations outside the text are numbered continuously by chapter and represented by two arabic numbers between parentheses and separated by a point (e.g. 3.5). Figures and tables are each numbered continuously by chapter and represented by two Arabic numbers (e.g. Figure 6.13 or Table 5.21). Bibliographic references are numbered continuously by chapter and represented by two Arabic numbers between square brackets [2.21].

1

NETWORK DESIGN

As soon as there are more than a few stations, they can no longer be interconnected with a simple multipoint line; it becomes necessary to organize exchanges on a network which operates a number of links interconnected by nodes (Figure 1.1). Such a structure poses many problems in comparison with a simple data link. At the design stage, it is convenient to determine the topology of the network in accordance with the constraints and criteria imposed on the designer. At the operating level, the network must be capable of routing messages and guaranteeing correct operation in spite of overloading and equipment failures.

The problems of network design will be tackled here; those connected with operation will be considered in the following chapter. The network consists of data links interconnected by intelligent nodes which are actually switches capable of temporarily storing incoming packets and routing them to one of the output lines (Figure 1.2). The packet switch is built around a processor which provides a large number of functions such as routing, control of buffer memories and line procedures together with packet concentration and multiplexing.

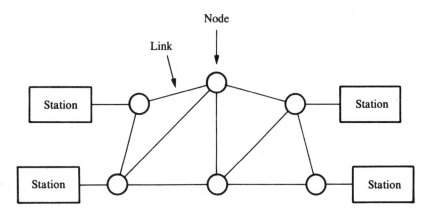

Figure 1.1 Network

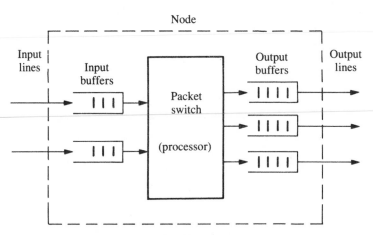

Figure 1.2 Node structure

The main problem for the network designer is to minimize the total cost of the system while guaranteeing a minimum performance in respect of message transfer time, availability and network capacity. The input data is the geographical location of the terminal stations, the mean traffic which they must exchange and the cost of the links. The designer can, within certain limits, vary the topology of the network, the capacity of the lines and the distribution of load between the various links in order to achieve the required optimum. The main tools used to achieve optimization of a network are graph theory, queuing theory and simulation. The theoretical aspects of network design will be undertaken here by recalling some essential elements of graphs and queues [1.1–1.35].

1.1 QUEUES

1.1.1 Introduction

In a teleprocessing network, messages are produced by stations in a random manner and the length of these messages is itself generally random. As the capacity of the lines is in principle fixed, messages which cannot be transmitted immediately must be placed in a *queue* and the mean *transfer time* $\bar{T}_R$ of a link by a message is given by

$$\bar{T}_R = \bar{T}_p + \bar{T}_t + \bar{T} \tag{1.1}$$

where $\bar{T}_p$, $\bar{T}_t$ and $\bar{T}$ are the *propagation*, *transmission* and *waiting times* respectively in the queue. When the traffic is high with respect to the capacity of the line, the waiting time $\bar{T}$ can be very much greater than the transmission and propagation times, and the performance of the network depends closely on the behaviour of the queues at the various nodes. In practice, the problem which arises is to determine the buffer memory sizes and to evaluate the waiting time as a function of the traffic flow and the capacity of the lines.

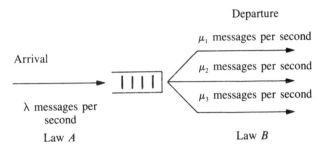

Figure 1.3 Queue

A queue (Figure 1.3) can be characterized by six parameters represented by the respective symbols A, B, C, K, m and Z as follows:

A **The input process**: The symbol A specifies the message input process law and thus defines the distribution of message arrivals.

B **The service process**: Messages are drawn through the queue by one or more *servers*, which in the case of a network are quite simply the output line or lines. The symbol B specifies the law which governs the service process.

C **The number of servers**: The number represented by C indicates the number of servers, in this case the number of output lines.

K **The maximum capacity of the queue**: The symbol K indicates the maximum number of messages which can be stored in the queue.

m **The population of users**.

Z **The service discipline**: The symbol Z specifies the manner in which the queue is managed, that is the order in which the arriving messages are arranged in the queue and the order in which the messages leave the queue. The most commonly used disciplines are FIFO (first in, first out), LIFO (last in, first out) and FIRO (first in, random out).

It is convenient to represent a queue by **Kendall's notation** which lists the six factors which characterize the queue as follows:

$$A/B/C/K/m/Z$$

Using this notation, C, K and m are replaced by the numbers which correspond to the particular queue considered, and Z is replaced by one of the symbols FIFO, LIFO, FIRO according to the service discipline. The letters A and B represent the input and output processes using the following symbols according to the law which is adopted:

D: Constant law
E_K: Erlang's law K
G: General law
GI: General independent law
H_K: Hyperexponential law of order K
M: Exponential law

In practice, most queues are represented only by the first three symbols of Kendall's notation. In this case, the last three symbols implicitly take the values $K = \infty$, $m = \infty$ and $Z = $ FIFO. Hence, the notation M/M/1 represent a one server queue with exponential distribution of arrivals and departures in which the service discipline is FIFO and the population of messages and the capacity of the queue are infinite. In view of the importance of the nature of the arrival and service laws in the study of queues, the main properties of the most common laws are briefly reviewed below.

1.1.2 Poisson law

The Poisson law is much used in queuing theory, since it corresponds well to the actual behaviour of a data processing system which contains a large number of users. The law is based on the following three assumptions:

(a) Arrivals of messages are independent of each other. This assumption is very well justified when the messages come from a large number of different random sources.
(b) The process is stationary, that is its characteristics do not vary with the start of the observation period.
(c) For a very short time interval Δt, the probability $\lambda \Delta t$ of the arrival of a single message is much greater than the probability of the arrival of several messages.

Starting from these assumptions, the probability $P_n(T)$ of the arrival of n messages during an interval T seconds will now be determined. The interval of T seconds can be divided into l intervals of duration Δt seconds where

$$T = l\Delta t \tag{1.2}$$

If the intervals Δt are sufficiently small, the third assumption implies that within each of these intervals only one message can arrive, with probability $\lambda \Delta t$, or no message arrives, with probability $1 - \lambda \Delta t$. The probability $P_n(l)$ that n messages arrive during the l intervals of duration Δt is then given by

$$P_n(l) = C_l^n (\lambda \Delta t)^n (1 - \lambda \Delta t)^{l-n} \tag{1.3}$$

where C_l^n is the number of combinations of n intervals from l. Then, by multiplying $(\lambda \Delta t)^n$ by $(l/l)^n$,

$$P_n(l) = \frac{l(l-1)\ldots[l-(n-1)](\lambda l\Delta t)^n(1-\lambda\Delta t)^{l-n}}{l^n n!} \tag{1.4}$$

or

$$P_n(l) = \left(1 - \frac{1}{l}\right)\ldots\left[1 - \frac{(n-1)}{l}\right]\frac{(\lambda l\Delta t)^n}{n!}(1-\lambda\Delta t)^l(1-\lambda\Delta t)^{-n} \tag{1.5}$$

When Δt tends to zero with $l\Delta t$ tending to T, $P_n(l)$ tends to $P_n(T)$ where

$$P_n(T) = \frac{(\lambda T)^n}{n!} \lim_{l \to \infty} \left(1 - \frac{\lambda T}{l}\right)^l \tag{1.6}$$

The limit of $(1 - \lambda T/l)^l$ is given by

$$\lim_{l \to \infty} \left(1 - \frac{\lambda T}{l}\right)^l = 1 - \lambda T + \frac{(\lambda T)^2}{2!} - \frac{(\lambda T)^3}{3!} + \ldots = e^{-\lambda T} \tag{1.7}$$

The probability $P_n(T)$ of the arrival of n messages during T seconds is then

$$P_n(T) = \frac{(\lambda T)^n e^{-\lambda T}}{n!} \tag{1.8}$$

This defines the *Poisson law*. It is possible to deduce the mean number $E_T(n)$ of message arrivals during T seconds immediately from this law:

$$E_T(n) = \sum_{n=0}^{\infty} n P_n(T) = e^{-\lambda T} \sum_{n=0}^{\infty} \frac{n(\lambda T)^n}{n!} = \lambda T e^{-\lambda T} \sum_{n=1}^{\infty} \frac{(\lambda T)^{n-1}}{(n-1)!} \tag{1.9}$$

or

$$E_T(n) = \lambda T \tag{1.10}$$

The Poisson process defined by (1.8) thus corresponds to a mean arrival of λ messages per second, which conforms to assumptions (a) and (b). The variance $V_T(n)$ of the number of message arrivals in T seconds is also obtained very simply as follows:

$$V_T(n) = \sum_{n=0}^{\infty} (n - \lambda T)^2 P_n(T) = \sum_{n=0}^{\infty} (n - \lambda T)^2 \frac{(\lambda T)^n e^{-\lambda T}}{n!} \tag{1.11}$$

$$V_T(n) = e^{-\lambda T} \left[(\lambda T)^2 \sum_{n=2}^{\infty} \frac{(\lambda T)^{n-2}}{(n-2)!} + \lambda T \sum_{n=1}^{\infty} \frac{(\lambda T)^{n-1}}{(n-1)!} \right.$$

$$\left. -2(\lambda T)^2 \sum_{n=1}^{\infty} \frac{(\lambda T)^{n-1}}{(n-1)!} + (\lambda T)^2 \sum_{n=0}^{\infty} \frac{(\lambda T)^n}{n!} \right] \tag{1.12}$$

$$V_T(n) = \lambda T \tag{1.13}$$

The variance is thus equal to the mean for a Poisson process.

Now consider the distribution of time intervals τ between successive arrivals of messages. If $f(\tau)$ is the probability density function of τ, the probability $f(\tau)d\tau$ that the time interval between two successive messages is equal to τ corresponds to the case where no message is received during time τ and a message arrives between times τ and $\tau + \Delta\tau$. Hence

$$f(\tau)d\tau = P_0(\tau) P_1(\Delta\tau) \tag{1.14}$$

and, using (1.8),

$$f(\tau)d\tau = \lambda e^{-\lambda \tau} d\tau \tag{1.15}$$

The probability density function of the intervals between arrivals is thus given by the **exponential law**

$$f(\tau) = \lambda e^{-\lambda\tau} \tag{1.16}$$

Under these conditions, the mean time $E(\tau)$ between arrivals can be obtained very simply from

$$E(\tau) = \int_0^\infty \tau f(\tau) d\tau = \int_0^\infty \lambda \tau e^{-\lambda\tau} d\tau \tag{1.17}$$

$$E(\tau) = \frac{1}{\lambda} \tag{1.18}$$

Similarly, the variance $V(\tau)$ of the time between arrivals is given by

$$V(\tau) = \int_0^\infty \left(\tau - \frac{1}{\lambda}\right)^2 \lambda e^{-\lambda\tau} d\tau \tag{1.19}$$

$$V(\tau) = \frac{1}{\lambda^2} \tag{1.20}$$

1.1.3 M/M/1 queue

The simplest queue will be considered here, it has a single server which in this case will be a line of capacity C bit/s (Figure 1.4). The queue is managed according to a FIFO discipline; the size of the buffer memory and message population are assumed to be infinite. Messages arrive in the queue following a Poisson distribution with a mean value of λ messages per second; it is assumed that the length of messages is exponentially distributed with a mean of $1/\mu$ bits per message. The probability density function $f(m)$ of the message length is thus given, using (1.16) and (1.18) by

$$f(m) = \mu e^{-\mu m} \tag{1.21}$$

As the output line has a fixed capacity of C bit/s, the assumption of an exponential distribution of message length also implies an exponential distribution of service times whose probability density function $f(t)$ is given by

$$f(t) = \mu C e^{-\mu C t} \tag{1.22}$$

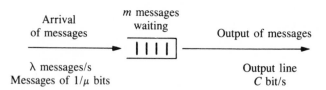

Figure 1.4 M/M/1 queue

with a mean value of $1/\mu C$. The arrival and service laws are thus exponential and this leads to adopting the notation M/M/1 to represent the queue. The choice of exponential laws to represent the arrival and service processes helps to simplify the calculations while ensuring a reasonable representation of actual system behaviour. It has been shown above that, when the messages are produced by a large number of independent sources, their arrival follows a Poisson law and the distribution of the time intervals between arrivals is exponential. This assumption is therefore close to reality. As far as the output process is concerned, the approximation of the model is much cruder, since the exponential service law implies that the length of messages is itself distributed according to an exponential law. This is strictly impossible since this law assumes that the messages can have any length while in practice they always have a length which is a multiple of a bit or a character. Furthermore, there is no a priori reason why the sources should produce messages whose length is exponentially distributed.

The M/M/1 model, therefore, allows only approximate treatment of the problems of queues in networks. Its principal use arises from the fact that exponential laws permit the mean number $\bar{m}$ of messages waiting in the queue and the corresponding mean waiting time $\bar{T}$ to be evaluated very simply. The assumption of an exponential distribution of arrivals and departures implies that, during a short time interval Δt, there can be only zero or one arrival and zero or one departure. Because of this, when the queue contains m messages at time t, it can contain only $m-1$, m or $m+1$ messages at time $t + \Delta t$ and its operation can be described by the state diagram of Figure 1.5 where the figures within the circles represent the number of messages waiting in the queue and the arcs defining the transitions are weighted by the probability of the corresponding transition.

Let $P_m(t + \Delta t)$ be the probability that there are m messages waiting at time $t + \Delta t$ and $P_{m-1}(t)$, $P_m(t)$ and $P_{m+1}(t)$ be the probabilities that there are $m-1$, m and $m+1$ messages respectively in the queue at time t. As there can be only one input and one output during the interval Δt, with respective transition probabilities of $\lambda\Delta t$ and $\mu C\Delta t$, the existence of m messages in the queue at time $t + \Delta t$ can arise only from the four possible states at time t represented in Table 1.1. Under these conditions, the probability $P_m(t + \Delta t)$ of having m messages in the queue at time $t + \Delta t$ is given by

$$P_m(t + \Delta t) = P_{m-1}(t)\lambda\Delta t(1 - \mu C\Delta t)$$
$$+ P_m(t)[(1 - \lambda\Delta t)(1 - \mu C\Delta t) + \lambda\Delta t\mu C\Delta t]$$
$$+ P_{m+1}(t)(1 - \lambda\Delta t)\mu C\Delta t \qquad (1.23)$$

or, neglecting terms in $(\Delta t)^2$

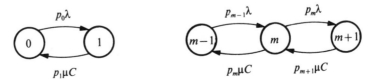

Figure 1.5 State diagram of an M/M/1 queue

Table 1.1 Possible changes of state of the queue between times t and $t + \Delta t$.

Number of messages in the queue at time t	Probability of the number of messages at time t	Inputs of messages	Outputs of messages	Probability of the corresponding event
$m - 1$	$p_{m-1}(t)$	1	0	$p_{m-1}(t)\lambda\Delta t(1 - \mu C\Delta t)$
m	$p_m(t)$	0	0	$p_m(t)(1 - \lambda\Delta t)(1 - \mu C\Delta t)$
m	$p_m(t)$	1	1	$p_m(t)\lambda\Delta t\mu C\Delta t$
$m + 1$	$p_{m+1}(t)$	0	1	$p_{m+1}(t)(1 - \lambda\Delta t)\mu C\Delta t$

$$\frac{P_m(t + \Delta t) - P_m(t)}{\Delta t} = \lambda P_{m-1}(t) - (\lambda + \mu C)P_m(t) + \mu C P_{m+1}(t) \quad (1.24)$$

Letting Δt tend to zero,

$$P'_m(t) = \lambda P_{m-1}(t) - (\lambda + \mu C)P_m(t) + \mu C P_{m+1}(t) \quad (1.25)$$

In the steady state, the derivatives $P'_m(t)$ of the state probabilities are zero and the state probabilities $P_m(t)$ tend to a value p_m which is independent of time. Under these conditions, (1.25) becomes

$$p_{m+1} = (1 + \rho)p_m - \rho p_{m-1} \quad (1.26)$$

with the **traffic intensity** ρ given by

$$\rho = \lambda/\mu C \quad (1.27)$$

By similar reasoning, it can easily be seen that

$$p_1 = \rho\, p_0 \quad (1.28)$$

Equations (1.26) and (1.28) enable p_m to be obtained by recurrence as a function of p_0:

$$p_2 = (1 + \rho)\, \rho\, p_0 - \rho\, p_0 = \rho^2 p_0$$

$$p_3 = (1 + \rho)\, \rho^2 p_0 - \rho^2 p_0 = \rho^3 p_0$$

$$p_m = \rho^m p_0 \quad (1.29)$$

p_0 can be evaluated very simply by noting that the sum of all the probabilities p_m is equal to 1:

$$\sum_{m=0}^{\infty} p_m = \sum_{m=0}^{\infty} \rho^m p_0 = \frac{p_0}{1 - \rho} = 1$$

$$p_0 = 1 - \rho \quad (1.30)$$

from which

$$p_m = (1 - \rho)\, \rho^m \quad (1.31)$$

The mean number $\bar{m}$ of messages waiting in the queue is thus

$$\bar{m} = \sum_{m=0}^{\infty} m p_m = \sum_{m=0}^{\infty} m(1-\rho)\rho^m \tag{1.32}$$

$$\bar{m} = \frac{\rho}{1-\rho} \tag{1.33}$$

The last equation indicates that, for an M/M/1 queue, the mean number of messages waiting in the queue depends only on one parameter, the traffic intensity ρ. When ρ tends to 1, $\bar{m}$ tends to infinity since the server can no longer dispose of messages at the rate at which they arrive, and they pile up in the queue. The mean number of waiting messages varies as a function of ρ, as shown in Figure 1.6.

The mean waiting time $\bar{T}$ in the queue is easily deduced from the mean number $\bar{m}$ of messages waiting, since the output line discharges on average μC messages per second. Hence:

$$\bar{T} = \frac{\bar{m}}{\mu C} = \frac{\rho}{\mu C(1-\rho)} \tag{1.34}$$

In practice, the user is generally interested in the mean transfer time $\bar{T}_R$ of the queue by a message. This time is the sum of the waiting time and the transmission time $1/\mu C$ of the output line where

$$\bar{T}_R = \frac{1}{\mu C} + \frac{\rho}{\mu C(1-\rho)} \tag{1.35}$$

$$\bar{T}_R = \frac{1}{\mu C(1-\rho)} = \frac{1}{\mu C - \lambda} \tag{1.36}$$

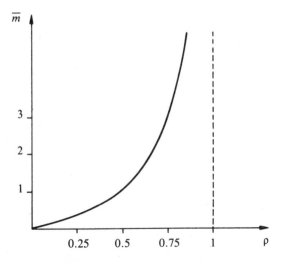

Figure 1.6 M/M/1 queue. Mean number of waiting messages as a function of traffic intensity

Comparison of Equations (1.33) and (1.36) shows that the mean transfer time and the mean number of messages are connected by the simple equation

$$\lambda \bar{T}_R = \bar{m} \tag{1.37}$$

This equation, which is called **Little's equation**, is quite general and applies to numerous cases of queues with much less restrictive assumptions than those relating to the definition of the M/M/1 queue. This equation indicates that, in a system of queues, the product of the mean rate of arrival of messages and the mean transfer time is equal to the mean number of messages waiting in the system. Little's equation will be used frequently in the following text.

1.1.4 M/M/1 queue with finite memory

So far it has been assumed that the storage capacity of the queue is infinite. In practice, this is clearly unacceptable since the messages waiting must be stored in a buffer memory whose size is necessarily finite. The problem which arises is to determine the buffer memory size and to evaluate the effect of memory limitation on the operation of the queue. According to Kendall's notation, an M/M/1 queue with a memory limited to a maximum of M messages should be denoted by M/M/1/M/∞/FIFO. For simplicity, this queue will still be considered to be M/M/1. A simple way to tackle the problem of the queue with finite memory consists of evaluating the probability $P(m > M)$ for which the number m of messages waiting in the queue exceeds M for the case of a queue with infinite memory. From (1.31):

$$P(m > M) = \sum_{m=M+1}^{\infty} p_m = \sum_{m=M+1}^{\infty} (1 - \rho)\rho^m \tag{1.38}$$

$$P(m > M) = \rho^{M+1} \tag{1.39}$$

This shows that for a traffic intensity of the order of 0.5, the overflow probability does not exceed 10^{-4}–10^{-5} for a buffer size of the order of 10–15 messages (Fig. 1.7).

The case of an M/M/1 queue with finite memory can be treated more rigorously by proceeding as for an infinite queue, but limiting the number of waiting messages to the maximum size M of the buffer. Under these conditions, (1.30) must be replaced by

$$\sum_{m=0}^{M} p_m = \sum_{m=0}^{M} \rho^m p_0 = p_0 \frac{1 - \rho^{M+1}}{1 - \rho} = 1 \tag{1.40}$$

which gives

$$p_0 = \frac{1 - \rho}{1 - \rho^{M+1}} \tag{1.41}$$

The probability p_m of having m messages waiting in the queue then becomes

$$p_m = \frac{(1 - \rho)\rho^m}{1 - \rho^{M+1}} \tag{1.42}$$

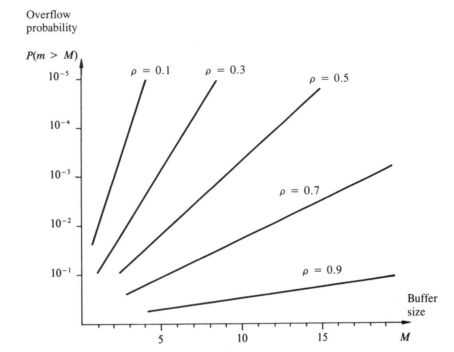

Figure 1.7 Overflow probability of an M/M/1 queue as a function of buffer size M and traffic intensity ρ

The blocking probability p_B corresponds to the case where the buffer is full and contains M messages. In this case

$$p_B = \frac{(1 - \rho)\rho^M}{1 - \rho^{M+1}} \tag{1.43}$$

1.1.5 M/G/1 queue

So far a queue with Poisson message arrivals and an exponential distribution of message length has been examined. It has been shown that the latter assumption is not very realistic since it does not permit the discrete nature of the messages to be taken into account and does not allow the distributions encountered in practice to be approached; in particular, the important case where the messages are of constant length is not accommodated. For a more realistic treatment of queues, it is desirable to retain the Poisson law for message arrival, since it is known that this law well reflects reality, but it is necessary to adopt a service time distribution which is better suited to real operating conditions. It would seem to be necessary to define a new model of the queue each time the service time distribution changes (recall that the law defining the distribution of message lengths is the same as that defining the service times, since the output line has a constant capacity C). Fortunately, this is not necessary since the operation of a queue can be described by a very general law, the **Pollaczek–Khinchine law**, which depends only on the traffic intensity, the mean message length and the variance of the message length. It is thus possible to define a model of the M/ G/1 queue for which the law of arrivals is Poisson and the service law, which is *general*, is defined only by its mean and its variance.

The behaviour of the queue can be examined by observing its operation between the departure of the message of index $i - 1$ and the departure of the message of index i. If the queue contains m_i messages after the departure of the message of index i, clearly

$$m_i = m_{i-1} + n_i - 1 \quad \text{for } m_{i-1} \geq 1 \tag{1.44}$$
$$= n_i \quad \text{for } m_{i-1} = 0$$

where n_i represents the number of messages which arrive in the queue during the random service time of the message of index i. The cases corresponding to $m_{i-1} = 0$ and $m_{i-1} \neq 0$ must be treated separately since there cannot be an output from the queue when it is empty, that is when $m_{i-1} = 0$. In practice, it is convenient to treat the two Equations (1.44) simultaneously using

$$m_i = m_{i-1} - U(m_{i-1}) + n_i \tag{1.45}$$

where $U(m_{i-1}) = 1$ for $m_{i-1} \neq 0$ and $U(m_{i-1}) = 0$ for $m_{i-1} = 0$. In the steady state, the number of messages waiting in the queue tends to a mean value $\bar{m}$, where

$$\bar{m}_i = \bar{m}_{i-1} = \bar{m} \tag{1.46}$$

Similarly, the number of messages n_i entering the queue during the service time tends to a mean value $\bar{n}$. Hence, by calculating the mean of the two terms of Equation (1.45)

$$\overline{U(m_{i-1})} = \bar{n} \tag{1.47}$$

The following is obtained by squaring the two terms of (1.45) and taking the mean:

$$\overline{m_i^2} = \overline{m_{i-1}^2} - 2\,\overline{m_{i-1}U(m_{i-1})} + 2\,\overline{m_{i-1}\,n_i} + \overline{U^2(m_{i-1})} - 2\,\overline{n_i\,U(m_{i-1})} + \overline{n_i^2} \tag{1.48}$$

Since $\overline{U^2(m_{i-1})} = \overline{U(m_{i-1})}$ and $\overline{m_{i-1}U(m_{i-1})} = \overline{m_{i-1}}$, using (1.46) and (1.47), Equation (1.48) becomes

$$\bar{m} = \frac{\bar{n}}{2} + \frac{V(n)}{2(1 - \bar{n})} \tag{1.49}$$

where $V(n)$ is the variance of the number of messages arriving during the random service time. If $\bar{\tau}$ is the mean service time, clearly

$$\bar{\tau} = \frac{1}{\mu C} \tag{1.50}$$

where $1/\mu$ is the mean message length expressed as a number of bits and C is the capacity of the output line expressed in bits per second. Little's formula implies that the mean number $\bar{n}$ of messages arriving during the service time is equal to the product of the mean number λ of messages arriving each second and the mean service time $\bar{\tau}$. Hence:

$$\bar{n} = \lambda\bar{\tau} = \frac{\lambda}{\mu C} = \rho \tag{1.51}$$

from which, using (1.49),

$$\bar{m} = \frac{\rho}{2} + \frac{V(n)}{2(1-\rho)} \tag{1.52}$$

So far no assumption has been made concerning the arrival and service laws. It will now be assumed that the arrivals are distributed according to a Poisson law and the service times τ are distributed with a probability density function $f(\tau)$. Under these conditions, the probability $P_K(\tau)$ of having K arrivals during τ seconds is given by

$$P_K(\tau) = \frac{(\lambda\tau)^K e^{-\lambda\tau}}{K!} \tag{1.53}$$

The probability P_K of having K message arrivals during a random service time is then

$$P_K = \int_0^\infty \frac{(\lambda\tau)^K e^{-\lambda\tau}}{K!} f(\tau) d\tau \tag{1.54}$$

The variance $V(n)$ of the number of arrivals can be determined from P_K using

$$V(n) = \sum_{K=0}^\infty \int_0^\infty (K-\rho)^2 \frac{(\lambda\tau)^K e^{-\lambda\tau}}{K!} f(\tau) d\tau \tag{1.55}$$

Reversing the order of summation and integration, this gives

$$V(n) = \int_0^\infty f(\tau) d\tau \left[\sum_{K=0}^\infty K^2 \frac{(\lambda\tau)^K e^{-\lambda\tau}}{K!} \right.$$

$$\left. - \sum_{K=0}^\infty 2K\rho \frac{(\lambda\tau)^K e^{-\lambda\tau}}{K!} + \sum_{K=0}^\infty \rho^2 \frac{(\lambda\tau)^K e^{-\lambda\tau}}{K!} \right] \tag{1.56}$$

$$V(n) = \lambda^2 \int_0^\infty \tau^2 f(\tau) d\tau + \lambda(1-2\rho) \int_0^\infty \tau f(\tau) d\tau + \rho^2 \int_0^\infty f(\tau) d\tau \tag{1.57}$$

Equation (1.57) shows that $V(n)$ can be expressed as a function of the first two moments of τ where

$$\int_0^\infty f(\tau) d\tau = 1 \tag{1.58}$$

$$\int_0^\infty \tau f(\tau) d\tau = \bar{\tau} = \rho/\lambda = 1/\mu C \tag{1.59}$$

$$\int_0^\infty \tau^2 f(\tau)d\tau = V(\tau) + \bar{\tau}^2 = V(\tau) + \frac{1}{\mu^2 C^2} \tag{1.60}$$

where $V(\tau)$ is the variance of the service time. Equation (1.57) then reduces to

$$V(n) = \lambda^2 V(\tau) + \rho \tag{1.61}$$

from which, using (1.52),

$$\bar{m} = \frac{1}{2(1-\rho)} \{2\rho - \rho^2 [1 - \mu^2 C^2 V(\tau)]\} \tag{1.62}$$

Equation (1.62) is the **Pollaczek–Khinchine** formula. It defines the mean number $\bar{m}$ of messages waiting in the queue as a function of only the traffic intensity ρ, the mean message length $1/\mu C$ normalized as a function of the capacity C of the output line and the variance $V(\tau)$ of the service time, or the message length which is equivalent. This formula is therefore of very general applicability.

The mean transfer time $\bar{T}_R$ of the queue is deduced very simply from the mean number $\bar{m}$ of waiting messages using Little's formula

$$\bar{T}_R = \bar{m}/\lambda \tag{1.63}$$

$$\bar{T}_R = \frac{1}{2\mu C(1-\rho)} \{2 - \rho[1 - \mu^2 C^2 V(\tau)]\} \tag{1.64}$$

In practice, it is often convenient to consider the mean waiting time $\bar{T}$ in the queue rather than the transfer time. Hence

$$\bar{T}_R = \bar{T} + \frac{1}{\mu C} = \bar{T} + \bar{\tau} \tag{1.65}$$

This leads to a very simple equation for $\bar{T}$ as a function of the second moment of τ, $\overline{\tau^2}$:

$$\bar{T} = \frac{\lambda \overline{\tau^2}}{2(1-\rho)} = \frac{\lambda \overline{\tau^2}}{2(1 - \lambda\bar{\tau})} \tag{1.66}$$

As a quick check, the Pollaczek–Khinchine equations will be applied to an M/M/1 queue. In this case, the service time distribution is exponential with a probability density function $f(\tau)$ given by

$$f(\tau) = \mu C e^{-\mu C \tau} \tag{1.67}$$

The variance of this distribution has been determined in (1.20) which in this case gives

$$V(\tau) = 1/\mu^2 C^2 \tag{1.68}$$

Substituting $V(\tau)$ as defined by (1.68) into (1.62) and (1.64), the mean number of waiting messages and the transfer time of the queue are obtained:

$$\bar{m} = \rho/(1-\rho) \tag{1.69}$$

$$\bar{T}_R = 1/\mu C(1-\rho) \tag{1.70}$$

These two values are identical to those which were determined directly in (1.33) and (1.36) for the M/M/1 queue.

Now consider the case of messages of constant length $1/\mu$. The variance $V(\tau)$ is zero so that $\bar{m}$ and $\bar{T}_R$ reduce to

$$\bar{m} = \frac{\rho}{(1 - \rho)}\left(1 - \frac{\rho}{2}\right) \tag{1.71}$$

$$\bar{T}_R = \frac{1}{\mu C(1 - \rho)}\left(1 - \frac{\rho}{2}\right) \tag{1.72}$$

This shows that, for the case where the messages have a constant length, the mean number of waiting messages and the transfer time are reduced by a factor $(1 - \rho/2)$ with respect to the case where the message lengths are exponentially distributed.

1.1.6 Geometric distribution

The M/G/1 queue model enables the cases which correspond to a Poisson arrival of messages whose length is distributed in any manner to be treated. In particular, this model resolves the problem of fixed-length messages. In practice, messages always have a length which is a discrete multiple of a word, which can be a bit, a character or a byte; continuous distributions, which have been examined so far, do not lead to a very realistic model except when the messages have a fixed length. Statistical studies have shown that, in practice, message lengths can often be considered to be distributed according to a **geometric law** which can be regarded as the discrete equivalent of the exponential law. If it is assumed that messages always have a length which is a multiple of a word, the geometric law defines the probability $P(l = L)$ that the message length L is equal to l given by

$$P(l = L) = p(1 - p)^{l-1} \tag{1.73}$$

where p is the probability that the message has a length equal to a word. Hence $p < 1$ and

$$\sum_{l=1}^{\infty} P(l = L) = \sum_{l=1}^{\infty} p(1 - p)^{l-1} = 1 \tag{1.74}$$

The mean message length $\bar{l}$ is given by

$$\bar{l} = \sum_{l=1}^{\infty} lp(1 - p)^{l-1} = 1/p \tag{1.75}$$

The variance $V(l)$ of the message length is defined by

$$V(l) = \sum_{l=1}^{\infty} \left(l - \frac{1}{p}\right)^2 p(1 - p)^{l-1}$$

$$V(l) = (1 - p)/p^2 \tag{1.76}$$

It can be seen that this geometric distribution tends to an exponential distribution of probability density $f(l)$ when p tends to zero. Consider the exponential distribution whose probability density $f(l)$ is given by

$$f(l) = p\,e^{-pl} \tag{1.77}$$

For $p \ll 1$,

$$f(0) = p$$
$$f(1) \simeq p(1 - p)$$
$$f(2) \simeq p(1 - 2p) \simeq p(1 - p)^2 \tag{1.78}$$

which clearly shows that the law is the same as for the geometric distribution defined by (1.73) when $p \ll 1$.

Finally notice that the case where the messages are of constant length can be derived from the geometric law by taking $p = 1$. Then $P(l = 1) = 1$ and $P(l \neq 1) = 0$, with $\bar{l} = 1$ and $V(l) = 0$.

1.1.7 Moment-generating function

Study of a discrete distribution can be greatly simplified by the use of moment-generating functions which can be defined as follows.

Consider a random variable l which represents, for example, the message length. This random variable can take the values 0, 1, 2, . . . , i, . . . , with respective probabilities of $p_0, p_1, p_2, \ldots, p_i, \ldots$. The **moment-generating function** $G(z)$ of this random variable is defined by

$$G(z) = \sum_{l=0}^{\infty} p_l z^l \tag{1.79}$$

It will be shown that the various moments of the distribution can be deduced very simply from successive derivatives of $G(z)$. Notice first that putting $z = 1$ gives

$$G(1) = \sum_{l=0}^{\infty} p_l = 1 \tag{1.80}$$

Differentiating $G(z)$ with respect to z:

$$\frac{dG(z)}{dz} = \sum_{l=1}^{\infty} l p_l z^{l-1} \tag{1.81}$$

This shows that by putting $z = 1$ in (1.81), the mean value $\bar{l}$ of l is obtained:

$$\frac{dG(z)}{dt}\bigg|_{z=1} = \sum_{l=1}^{\infty} l p_l = \bar{l} \tag{1.82}$$

Now consider the second derivative of $G(z)$

$$\frac{\mathrm{d}^2 G(z)}{\mathrm{d}z^2} = \sum_{l=2}^{\infty} l(l-1)p_l z^{l-2} = \sum_{l=2}^{\infty} l^2 p_l z^{l-2} - \sum_{l=2}^{\infty} l p_l z^{l-2} \qquad (1.83)$$

$$\left.\frac{\mathrm{d}^2 G(z)}{\mathrm{d}z^2}\right|_{z=1} = \sum_{l=2}^{\infty} l^2 p_l - \sum_{l=2}^{\infty} l p_l + p_1 - p_1 \qquad (1.84)$$

$$\left.\frac{\mathrm{d}^2 G(z)}{\mathrm{d}z^2}\right|_{z=1} = \overline{l^2} - \overline{l} \qquad (1.85)$$

In a similar manner, the higher-order derivatives allow the various moments of the random variable to be obtained.

The use of moment-generating functions considerably simplifies the problems posed by the sums of independent random variables. If a second random variable k, which is independent of l, is considered, its moment-generating function $H(z)$ is defined by

$$H(z) = \sum_{k=0}^{\infty} q_k z^k \qquad (1.86)$$

where $q_0, q_1, q_2, \ldots, q_i, \ldots$ are the probabilities that k takes the values 0, 1, 2, ..., i, ... The probabilities that the sum of the two random variables takes the values 0, 1, 2, ... are then $p_0 q_0$, $p_0 q_1 + p_1 q_0$, $p_0 q_2 + p_1 q_1 + p_2 q_0$, ... which indicates that the generating function $A(z)$ of the sum of the two random variables l and k is given by

$$A(z) = H(z)G(z) \qquad (1.87)$$

It can be seen that the generating function of the sum of two independent random variables is equal to the product of the generating functions of the two random variables. This property is very useful and will be used in the following section.

1.1.8 Queue with Poisson arrivals and geometric distribution of message length

To analyse a queue with Poisson message arrivals and a geometric distribution of message length, the procedure is very similar to that adopted in § 1.1.5 for the M/G/1 queue, but the evolution of the state of the queue during a constant service time Δt will be considered instead of a random service time. It will be assumed that the output line has a capacity of C words per second, which implies that a word leaves the queue during each time interval $\Delta t = 1/C$, except when the queue is empty (Figure 1.8). Under these conditions, if n_i is the number of words entering the queue during the interval Δt, and m_{i-1} and m_i are the number of words in the queue at the beginning and end of a time interval Δt respectively, the following relation exists:

$$m_i = m_{i-1} - U(m_{i-1}) + n_i \qquad (1.88)$$

where $U(m_{i-1}) = 1$ for $m_{i-1} \neq 0$ and $U(m_{i-1}) = 0$ for $m_{i-1} = 0$. By taking the mean of the two terms of (1.88), a steady-state solution is obtained:

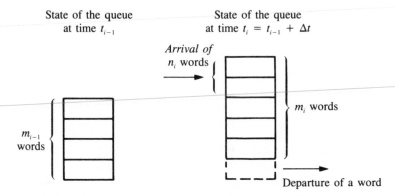

Figure 1.8 Evolution of the state of the queue during the service time Δt

$$\overline{U(m_{i-1})} = \bar{n} \tag{1.89}$$

where $\bar{n}$ is the mean number of words entering the queue during the service time of a word $\Delta t = 1/C$. The mean number $\bar{m}$ of words waiting in the queue can be obtained simply by squaring the two terms of (1.88) and taking the mean:

$$\overline{m_i^2} = \overline{m_{i-1}^2} - \overline{2m_{i-1}U(m_{i-1})} + \overline{2m_{i-1}n_i} + \overline{U^2(m_{i-1})} - \overline{2n_iU(m_{i-1})} + \overline{n_i^2} \tag{1.90}$$

from which the following equation is obtained:

$$\bar{m} = \frac{\bar{n}}{2} + \frac{V(n)}{2(1-\bar{n})} \tag{1.91}$$

This is identical to (1.49) but here $V(n)$ is the variance of the number of words entering the queue during the time interval $\Delta t = 1/C$.

To determine $\bar{m}$, it is necessary to know $\bar{n}$ and $V(n)$ which is possible by using the laws which determine the distribution of arrivals and message lengths. Messages arrive in the queue according to a Poisson law with a mean number λ of messages per second. The probability q_K of the arrival of K messages during $1/C$ is thus given by

$$q_K = \frac{(\lambda/C)^K e^{-\lambda/C}}{K!} \tag{1.92}$$

The length l of the messages which arrive in the queue is defined by a geometric distribution of mean value $\bar{l} = 1/p$. The probability p_l that a message entering the queue has a length of l words is thus

$$p_l = \frac{1}{\bar{l}}\left(1 - \frac{1}{\bar{l}}\right)^{l-1} \tag{1.93}$$

The moment-generating function $G(z)$ of the geometric distribution is given by

$$G(z) = pz + p(1-p)z^2 + p(1-p)^2z^3 + \dots \tag{1.94}$$

$$G(z) = pz\left[1 + (1-p)z + (1-p)^2z^2 + \dots\right] \tag{1.95}$$

$$G(z) = \frac{pz}{1 - (1 - p)z} \tag{1.96}$$

The mean number $\bar{m}$ of words waiting in the queue will now be found using (1.91). To do this, it is necessary to determine the mean number $\bar{n}$ and the variance $V(n)$ of the number of words entering the queue during the interval Δt. These two values can be obtained very simply by means of the moment-generating function $H(z)$ of n. During the interval Δt, 0, 1, 2, . . . , K, . . . messages can arrive with respective probabilities of $q_0, q_1, q_2, \ldots, q_K, \ldots$ where q_K is given by (1.92). When K messages arrive in the queue, the total number of words which enter the queue is the sum of K random variables defined by the geometric distribution (1.93). The corresponding moment-generating function is thus equal to the product of K moment-generating functions $G(z)$ of the geometric distribution, since a sum of random variables can be represented by the product of their generating functions, as shown in the previous section. The generating function $H(z)$ of n is thus defined by

$$H(z) = \sum_{K=0}^{\infty} q_K (G(z))^K \tag{1.97}$$

and, using (1.92) and (1.96),

$$H(z) = \sum_{K=0}^{\infty} \frac{(\lambda/C)^K e^{-\lambda/C}(pz)^K}{K! [1 - (1 - p)z]^K} \tag{1.98}$$

As a quick check, notice that $H(1) = 1$ which conforms to the fundamental property of moment-generating functions. Summing the series (1.98), $H(z)$ becomes

$$H(z) = e^{-\lambda/C} e^{\lambda pz/\{C[1 - (1 - p)z]\}} \tag{1.99}$$

The derivative of $H(z)$ is given by

$$\frac{dH(z)}{dz} = e^{-\lambda/C} e^{\lambda pz/\{C[1 - (1 - p)z]\}} \left[\frac{1 - (1 - p)z + (1 - p)z}{C[1 - (1 - p)z]^2} \right] \lambda p \tag{1.100}$$

$$\frac{dH(z)}{dz} = e^{-\lambda/C} e^{\lambda pz/\{C[1 - (1 - p)z]\}} \frac{\lambda p}{C[1 - (1 - p)z]^2} \tag{1.101}$$

This enables the mean number $\bar{n}$ of words which arrive in the queue during the interval Δt to be obtained:

$$\bar{n} = \left. \frac{dH(z)}{dz} \right|_{z=1} = \frac{\lambda}{pC} = \frac{\lambda \bar{l}}{C} \tag{1.102}$$

The variance $V(n)$ can now be evaluated from the second-order moment by calculating the second derivative of $H(z)$ and using Equation (1.85):

$$\frac{d^2 H(z)}{dz^2} = e^{-\lambda/C} e^{\lambda pz/\{C[1 - (1 - p)z]\}} \frac{\lambda p}{C} \left[\frac{\lambda p}{C[1 - (1 - p)z]^4} \right.$$

$$\left. + \frac{2(1 - p)}{[1 - (1 - p)z]^3} \right] \tag{1.103}$$

$$\overline{n^2} - \overline{n} = \frac{d^2H(z)}{dz^2}\bigg|_{z=1} = \frac{\lambda}{Cp^2}\left[\frac{\lambda}{C} + 2(1-p)\right] \qquad (1.104)$$

Using $V(n) = \overline{n^2} - \overline{n}^2$ and (1.102), the variance $V(n)$ becomes

$$V(n) = \frac{\lambda}{Cp^2}(2-p) = \frac{\lambda \overline{l}^2}{C}\left(2 - \frac{1}{\overline{l}}\right) \qquad (1.105)$$

This enables the mean number $\overline{m}$ of words waiting in the queue to be obtained by substituting $\overline{n}$ and $V(n)$, defined by (1.102) and (1.105) respectively, into (1.91)

$$\overline{m} = \frac{\lambda}{2pC} + \frac{\lambda(2-p)}{2p(pC - \lambda)} \qquad (1.106)$$

$$\overline{m} = \frac{\lambda \overline{l}}{2C} + \frac{\overline{l}\lambda(2\overline{l} - 1)}{2(C - \lambda \overline{l})} \qquad (1.107)$$

The mean transfer time $\overline{T}_R$ of the queue by a message is derived from $\overline{m}$, using Little's formula, by noting from (1.102) that the mean number of arrivals of words per second is equal to $\lambda \overline{l}$. Hence

$$\overline{T}_R = \overline{m}/\lambda \overline{l} \qquad (1.108)$$

$$\overline{T}_R = \frac{1}{2C} + \frac{2\overline{l} - 1}{2(C - \lambda \overline{l})} \qquad (1.109)$$

It was shown in §1.1.6 that the case where the messages have a constant length corresponds to a geometric distribution with $p = 1$. With this assumption, Equations (1.107) and (1.109) reduce to

$$\overline{m} = \frac{\lambda/C}{1 - \lambda/C}\left(1 - \frac{\lambda}{2C}\right) \qquad (1.110)$$

$$\overline{T}_R = \frac{1}{C(1 - \lambda/C)}\left(1 - \frac{\lambda}{2C}\right) \qquad (1.111)$$

These equations can be compared with those obtained in (1.71) and (1.72) from the model of the M/G/1 queue. However, as the messages here are of constant length and equal to 1, C represents the number of messages per second absorbed by the output line, while, in the case corresponding to (1.71) and (1.72), C represents the capacity of the output line expressed in bits per second, and the length of the messages is $1/\mu$. It can be seen that, in the case corresponding to (1.71) and (1.72), the capacity of the output line is equal to μC messages per second. Under these conditions, it is easy to show that by replacing C by μC in (1.110) and (1.111) and with $\rho = \lambda/\mu C$, (1.71) and (1.72) are obtained again, which confirms the approach used here.

Returning to the general case which corresponds to a geometric distribution of message length, it is possible to modify Equations (1.107) and (1.109) by noting that $\lambda \overline{l}$ represents the mean number of words which enter the queue per second and C represents the number of words which leave the queue per second. The traffic intensity ρ is now given by

$$\rho = \frac{\lambda \bar{l}}{C} \tag{1.112}$$

This leads to replacing (1.107) and (1.109) by

$$\bar{m} = \frac{\rho}{2} + \frac{\rho(2\bar{l} - 1)}{2(1 - \rho)} \tag{1.113}$$

$$\bar{T}_R = \frac{1}{2C} + \frac{2\bar{l} - 1}{2C(1 - \rho)} \tag{1.114}$$

with

$$V(n) = \rho(2\bar{l} - 1) \tag{1.115}$$

Equations (1.113)–(1.115) clearly show that for the same traffic intensity ρ, the variance $V(n)$ of the number of words entering the queue, together with the mean number $\bar{m}$ of waiting words and the mean transfer time $\bar{T}_R$, increases as a function of the length of the messages. In other words, *for the same throughput of words in the queue, the transfer time and the length of the queue increase as the message lengths increase.* This behaviour of the queue with a geometric distribution of message length is illustrated in Figure 1.9 which shows the number of words waiting in the queue as a function of the traffic intensity and the mean

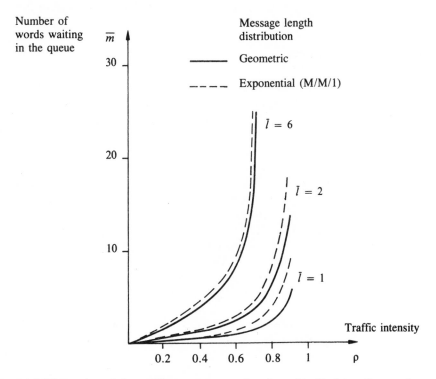

Figure 1.9 Mean size of the waiting queue as a function of traffic intensity ρ and mean message length $\bar{l}$

message length. It should be noted that the curve corresponding to $\bar{l} = 1$ represents the case where the messages are of constant length and this condition leads to a minimum size of the queue for a given traffic intensity.

These results can be compared with those given by the M/M/1 model which correspond to an exponential distribution of message length. It was seen in (1.33) that the mean number of waiting messages is, in this case, $\rho/(1 - \rho)$. As the messages have a mean length of $\bar{l}$ words, the mean number $\bar{m}_1$ of words waiting in an M/M/1 queue is defined by

$$\bar{m}_1 = \frac{\rho\bar{l}}{(1 - \rho)} \tag{1.116}$$

It can be seen that

$$\bar{m} = \bar{m}_1 - \frac{\rho^2}{2(1 - \rho)} \tag{1.117}$$

which indicates that the M/M/1 model always gives a queue size greater than that which corresponds to a geometric distribution of message lengths (Figure 1.9). It can be seen that the results obtained with a geometric distribution are between those given by the M/M/1 model and the model with constant length messages, since the latter always gives smaller queue sizes than with a geometric distribution. Notice also that when the mean message length $\bar{l}$ becomes large, $\rho^2/[2(1 - \rho)]$ becomes negligible in comparison with $\bar{m}$ and m_1 so that the results obtained with geometric and exponential distributions tend to be the same.

1.1.9 Queues with priorities

So far it has been assumed that the messages are of the same type and are served in the queue with the same priority. In practice it often happens that the messages handled by the network correspond to different classes of service and must consequently be treated by the various nodes in accordance with their level of priority. The most frequent case corresponds to a network where text messages and service or acknowledgement messages coexist and the latter are shorter and more urgent than text messages. The behaviour of a queue which processes r classes of messages whose respective priorities pr are $1, 2, \ldots, r$ will be examined; it is assumed here that the messages with the lowest priority index are served first. Messages belonging to the class of priority i have a mean length of $1/\mu_i$ bits and arrive in the queue at the rate of λ_i messages per second. The mean waiting time $\bar{T}_{pr}$ of messages of priority pr will be determined. If there are m_K messages in the queue of priority K equal to or less than pr, the waiting time $\bar{T}_{pr}$ is equal to the sum of the mean service times $\bar{T}_K^1$ of the m_K messages for $K = 1, 2, \ldots, pr$, to which must be added the sum of the mean service times $\bar{T}_K^2$ of the messages of priority less than pr which arrive during the interval $\bar{T}_{pr}$, together with the mean service time $\bar{T}_0$ of the message in the process of leaving the queue. Hence

$$\bar{T}_{pr} = \sum_{K=1}^{pr} \bar{T}_K^1 + \sum_{K=1}^{pr-1} \bar{T}_K^2 + \bar{T}_0 \tag{1.118}$$

If C is the capacity of the output line, expressed in bits per second, the mean service time $\bar{T}_K^1$ of the $\bar{m}_K$ messages belonging to class K which are waiting in the queue is given by

$$\bar{T}_K^1 = \bar{m}_K / \mu_K C \tag{1.119}$$

The waiting time $\bar{T}_K$ of the $\bar{m}_K$ messages in the queue is defined by

$$\lambda_K \bar{T}_K = \bar{m}_K \tag{1.120}$$

Hence

$$\bar{T}_K^1 = \rho_K \bar{T}_K \tag{1.121}$$

with

$$\rho_K = \lambda_K / \mu_K C \tag{1.122}$$

On average $\lambda_K \bar{T}_{pr}$ messages of class K arrive in the queue during the mean waiting time $\bar{T}_{pr}$ of the messages of priority pr. These messages of class K are served during a period $\bar{T}_K^2$ with a throughput $\mu_K C$. Hence

$$\bar{T}_K^2 = \rho_K \bar{T}_{pr} \tag{1.123}$$

Under these conditions, Equation (1.118) can be expressed as a function of the waiting times of the various classes, using equations (1.121) and (1.123), by

$$\bar{T}_{pr} = \sum_{K=1}^{pr} \rho_K \bar{T}_K + \bar{T}_{pr} \sum_{K=1}^{pr-1} \rho_K + \bar{T}_0 \tag{1.124}$$

where

$$\sum_{K=1}^{pr-1} \rho_K = 0 \text{ for } K = 1 \tag{1.125}$$

Equation (1.124) can be modified to form a recurrence relation with

$$\bar{T}_{pr}\left(1 - \sum_{K=1}^{pr-1} \rho_K\right) = \sum_{K=1}^{pr} \rho_K \bar{T}_K + \bar{T}_0 \tag{1.126}$$

$$\bar{T}_{pr}\left(1 - \sum_{K=1}^{pr} \rho_K\right) = \sum_{K=1}^{pr-1} \rho_K \bar{T}_K + \bar{T}_0 \tag{1.127}$$

$$\bar{T}_{pr}\left(1 - \sum_{K=1}^{pr} \rho_K\right) = \bar{T}_{pr-1}\left(1 - \sum_{K=1}^{pr-2} \rho_K\right) \tag{1.128}$$

from which, by multiplying the two terms by the same quantity,

$$\bar{T}_{pr}\left(1 - \sum_{K=1}^{pr} \rho_K\right)\left(1 - \sum_{K=1}^{pr-1} \rho_K\right) = \bar{T}_{pr-1}\left(1 - \sum_{K=1}^{pr-1} \rho_K\right)\left(1 - \sum_{K=1}^{pr-2} \rho_K\right) \tag{1.129}$$

The following is obtained by proceeding stepwise from $\bar{T}_1(1 - \rho_1) = \bar{T}_0$:

$$\bar{T}_{pr} = \frac{\bar{T}_0}{\left(1 - \sum_{K=1}^{pr} \rho_K\right)\left(1 - \sum_{K=1}^{pr-1} \rho_K\right)} \tag{1.130}$$

Consider the case of an M/G/1 queue without priorities. Equation (1.130) then reduces to

$$\bar{T}_{pr} = \frac{\bar{T}_0}{(1 - \rho)} \qquad (1.131)$$

By comparison with (1.66), it can be seen that

$$\bar{T}_0 = \frac{\lambda}{2} \overline{\tau^2} \qquad (1.132)$$

where τ is the duration of a random service interval. In a similar manner, the M/G/1 queue with priorities can be defined using

$$\bar{T}_0 = \frac{1}{2} \sum_{K=1}^{r} \lambda_K \overline{\tau^2} \qquad (1.133)$$

from which

$$\bar{T}_{pr} = \frac{\displaystyle\sum_{K=1}^{r} \lambda_K \overline{\tau_K^2}}{2\left(1 - \displaystyle\sum_{K=1}^{pr} \rho_K\right)\left(1 - \displaystyle\sum_{K=1}^{pr-1} \rho_K\right)} \qquad (1.134)$$

This equation generalizes the Pollaczek–Khinchine formula to the case of a queue with priorities. When the length of the messages is distributed exponentially, the queue reduces to the M/M/1 model with priorities. In this case, $\overline{\tau^2}$ becomes, using (1.20),

$$\overline{\tau^2} = 2/\mu_K^2 C^2 \qquad (1.135)$$

and the mean waiting time $\bar{T}_{pr}$ corresponding to the priority pr becomes

$$\bar{T}_{pr} = \frac{\displaystyle\sum_{K=1}^{r} \rho_K/\mu_K C}{\left(1 - \displaystyle\sum_{K=1}^{pr} \rho_K\right)\left(1 - \displaystyle\sum_{K=1}^{pr-1} \rho_K\right)} \qquad (1.136)$$

Comparison with Equation (1.34) shows that the waiting time corresponding to the conventional M/M/1 queue is obtained when $r = 1$. The outcome of introducing priorities into the queue operation can be analysed simply in the case where there are only two priority levels 1 and 2. Then, for level 1 messages, that is the most urgent,

$$\bar{T}_1 = \frac{(\rho_1/\mu_1 C) + (\rho_2/\mu_2 C)}{1 - \rho_1} \qquad (1.137)$$

and, for level 2 messages, that is the least urgent,

$$\bar{T}_2 = \frac{(\rho_1/\mu_1 C) + (\rho_2/\mu_2 C)}{(1 - \rho_1)(1 - \rho_1 - \rho_2)} \qquad (1.138)$$

from which

$$\bar{T}_2 = \frac{\bar{T}_1}{(1 - \rho_1 - \rho_2)} \qquad (1.139)$$

This indicates that the introduction of priorities slows down the flow of low priority messages since $\bar{T}_2 > \bar{T}_1$.

If the messages were transmitted without priority, the mean waiting time would become

$$\bar{T}_3 = \frac{(\rho_1/\mu_1 C) + (\rho_2/\mu_2 C)}{(1 - \rho_1 - \rho_2)} \qquad (1.140)$$

It can be seen, by comparison with (1.138), that the delay of messages of priority 2 is only slightly greater than that which would arise if there were no priorities, on condition that $1 - \rho_1$ is not much different from 1, that is on condition that the traffic intensity ρ_1 of the high priority messages is small. This explains why many networks use a system of priorities for transmission of urgent messages.

1.1.10 Open networks of queues

Teleprocessing networks consist of nodes which are interconnected by communication lines. Each node is equipped with a communication controller which routes the incoming messages to the queues for their output lines. A teleprocessing network can thus be considered as a network of queues. These networks are described as *open* when there is an exchange of messages with the outside world and *closed* otherwise. Analysis of networks of queues is in general difficult, and attention here will be restricted to those which are the simplest and which are based on a Poisson message arrival and an exponential distribution of service times [1.11]–[1.15].

The most elementary model will be considered first; this is an open network consisting of two queues connected in series with a Poisson arrival of messages, characterized by a mean value λ of messages per second and with an exponential distribution of message lengths of mean value $1/\mu$ bits per message (Figure 1.10). The output lines of queues Q1 and Q2 have respective capacities of C_1 and C_2 bits per second. Let $P_{t+\Delta t}(m_1,m_2)$ be the probability that the queues Q1 and Q2 contain m_1 and m_2 messages respectively at time $t + \Delta t$; $P_{t+\Delta t}(m_1,m_2)$ will now be expressed as a function of the various possible states of the queues at time t. As the arrivals are Poisson, one message at most can enter the queue during the interval Δt if the latter is chosen to be sufficiently small. The probability of the occurrence of this event is $\lambda \Delta t$ and the probability of the converse is $1 - \lambda \Delta t$. Similarly, the queues can at most lose only one message during Δt, with respective probabilities of $\mu C_1 \Delta t$ and $\mu C_2 \Delta t$. Notice that if the queue Q1 loses a message, this must enter the queue Q2 at the same time. Under these conditions, state (m_1,m_2) can be attained at the end of the time Δt only in the six cases indicated in Table 1.2, and $P_{t+\Delta t}(m_1,m_2)$ is given by

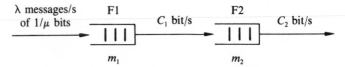

Figure 1.10 Network of two waiting queues in series

Table 1.2 Table of possible transitions affecting the state (m_1, m_2) after time Δt.

Number of messages in the queues at time t		Queue Q1		Queue Q2		Probability of transition to state (m_1, m_2)
Q1	Q2	Inputs of messages	Outputs of messages	Inputs of messages	Outputs of messages	
$m_1 - 1$	m_2	1	0	0	0	$\lambda \Delta t(1 - \mu C_1 \Delta t)(1 - \mu C_2 \Delta t)$
$m_1 - 1$	$m_2 + 1$	1	0	0	1	$\lambda \Delta t(1 - \mu C_1 \Delta t)\mu C_2 \Delta t$
m_1	m_2	0	0	0	0	$(1 - \lambda \Delta t)(1 - \mu C_1 \Delta t)(1 - \mu C_2 \Delta t)$
m_1	m_2	1	1	1	1	$\lambda \Delta t \mu C_1 \Delta t \mu C_2 \Delta t$
m_1	$m_2 + 1$	0	0	0	1	$(1 - \lambda \Delta t)(1 - \mu C_1 \Delta t)\mu C_2 \Delta t$
$m_1 + 1$	$m_2 - 1$	0	1	1	0	$(1 - \lambda \Delta t)\mu C_1 \Delta t(1 - \mu C_2 \Delta t)$

$$
\begin{aligned}
P_{t+\Delta t}(m_1, m_2) = {} & P_t(m_1 - 1, m_2)\lambda \Delta t(1 - \mu C_1 \Delta t)(1 - \mu C_2 \Delta t) \\
& + P_t(m_1 - 1, m_2 + 1)\lambda \Delta t(1 - \mu C_1 \Delta t)\mu C_2 \Delta t \\
& + P_t(m_1, m_2)(1 - \lambda \Delta t)(1 - \mu C_1 \Delta t)(1 - \mu C_2 \Delta t) \\
& + P_t(m_1, m_2)\lambda \Delta t\, \mu C_1 \Delta t\, \mu C_2 \Delta t \\
& + P_t(m_1, m_2 + 1)(1 - \lambda \Delta t)\mu C_2 \Delta t(1 - \mu C_1 \Delta t) \\
& + P_t(m_1 + 1, m_2 - 1)(1 - \lambda \Delta t)\mu C_1 \Delta t(1 - \mu C_2 \Delta t) \quad (1.141)
\end{aligned}
$$

Letting Δt tend to zero gives

$$
\begin{aligned}
P'_t(m_1, m_2) = {} & \lambda P_t(m_1 - 1, m_2) - (\lambda + \mu C_1 + \mu C_2)P_t(m_1, m_2) \quad (1.142) \\
& + \mu C_2 P_t(m_1, m_2 + 1) + \mu C_1 P_t(m_1 + 1, m_2 - 1)
\end{aligned}
$$

In the steady state, $P'_t(m_1, m_2)$ is zero and the system state equation reduces to

$$
\begin{aligned}
(\lambda + \mu C_1 + \mu C_2)P(m_1, m_2) = {} & \lambda P(m_1 - 1, m_2) \\
& + \mu C_2 P(m_1, m_2 + 1) \quad (1.143) \\
& + \mu C_1 P(m_1 + 1, m_2 - 1)
\end{aligned}
$$

where $P(m_1, m_2)$ is the steady-state probability of state (m_1, m_2). In a similar manner, the initial conditions are defined by

$$
\lambda P(0, 0) = \mu C_2 P(0, 1) \quad (1.144)
$$

$$
(\lambda + \mu C_2)P(0, m_2) = \mu C_1 P(1, m_2 - 1) + \mu C_2 P(0, m_2 + 1) \quad (1.145)
$$

$$
(\lambda + \mu C_1)P(m_1, 0) = \mu C_2 P(m_1, 1) + \lambda P(m_1 - 1, 0) \quad (1.146)
$$

The set of Equations (1.143)–(1.146) is solved in a similar manner to that of § 1.1.3 for the case of the M/M/1 queue and the final result is

$$P(m_1, m_2) = \rho_1^{m_1} \rho_2^{m_2} (1 - \rho_1)(1 - \rho_2) \tag{1.147}$$

where

$$\rho_1 = \lambda/\mu C_1, \qquad \rho_2 = \lambda/\mu C_2 \tag{1.148}$$

This shows that the state of the network can be analysed by considering each queue separately. The results obtained in (1.31) show that the probabilities $P(m_1)$ and $P(m_2)$ of having m_1 and m_2 messages waiting in the two M/M/1 queues are $(1 - \rho_1)\rho_1^{m_1}$ and $(1 - \rho_2)\rho_2^{m_2}$. Hence

$$P(m_1, m_2) = P(m_1)P(m_2) \tag{1.149}$$

which implies that the two queues are independent and queue Q2 can be treated with an M/M/1 model. This result is a consequence of the fact that *the output process of an M/M/1 queue is Poisson* and it can be generalized to a network of queues without feedback of the type represented in Figure 1.11 with

$$P(m_1, m_2, \ldots, m_N) = P(m_1)P(m_2) \ldots P(m_N) \tag{1.150}$$

When the network contains feedback loops, as in Figure 1.12, it can be shown that the output process of the queues is no longer Poisson because of the correlations introduced by the feedback. However, even in this case, the state probability remains equal to the product of the state probabilities of the various nodes as indicated in (1.150).

Under these conditions, if the network contains N data links, that is N queues, the analysis is performed in three stages:

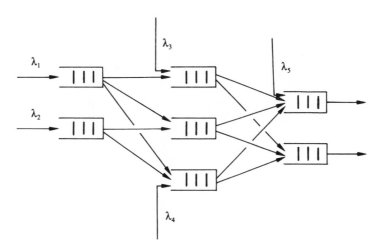

Figure 1.11 Example of an open network without feedback

Step 1

In the first step, the designer determines the various routes between the source and destination nodes. This step will be examined in the remainder of this chapter and the following one. Attention will not be devoted here to the manner in which the routes are chosen and it will be simply assumed that the network exploits R routes to ensure traffic flow between R pairs of source–destination nodes which are liable to communicate. Each route is described simply by the list of queues which are passed through, for example 1, 5, 9 for the route connecting nodes 1 and 7 of Figure 1.12.

Step 2

After the first step, the routes which carry the traffic between each source–destination pair are known, together with the traffic $\lambda^{(r)}$ on each route which is part of the specification. It is thus possible to compute the total traffic λ_i entering each queue i by calculating the sum of the traffic which flows through this queue:

$$\lambda_i = \sum_{r \in i} \lambda^{(r)}, \quad r = 1, \ldots, R, \ i = 1, \ldots, N \qquad (1.151)$$

Step 3

The mean number $\bar{m}_i^{(r)}$ of messages on route r while waiting in queue i is calculated using the M/M/1 model with

$$\bar{m}_i^{(r)} = \frac{\rho_i^{(r)}}{1 - \rho_i} \qquad (1.152)$$

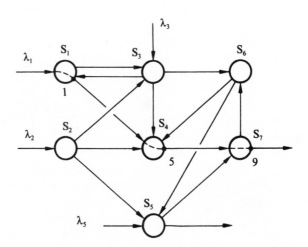

Figure 1.12 Example of an open network with feedback

$$\rho_i^{(r)} = \lambda^{(r)}/\mu C_i \tag{1.153}$$

$$\rho_i = \lambda_i/\mu C_i \tag{1.154}$$

where C_i is the capacity of the output line of queue i and $1/\mu$ is the mean message length. The total number $\bar{m}_i$ of messages waiting in queue i is clearly the sum of the number of messages corresponding to the various routes passing through the queue with

$$\bar{m}_i = \frac{\rho_i}{1 - \rho_i} \tag{1.155}$$

The mean transfer time $\bar{T}_R^{(r)}$ on route r is obtained by forming the sum of the transfer times along route r obtained from (1.152) using Little's formula

$$\bar{T}_R^{(r)} = \sum_{i \in r} \frac{1}{\mu C_i(1 - \rho_i)} \tag{1.156}$$

It can be seen that the model allows the computation of the various parameters of the network. The validity of such a model can seem dubious, since the M/M/1 queue assumes the independence of the random variables representing the times of arrival and the service times; this cannot be the case along the same route since the packets have the same length throughout the route. However, Kleinrock has shown that this condition of independence is in general preserved when the messages arrive in the queue from several different sources, and this justifies the use of this model.

1.1.11 Closed queuing networks

In practice, the size of the buffer memories in the network queues is always finite. In the case of a simple M/G/1 queue, limitation of the number of waiting messages to a maximum value M leads either to discarding the messages which arrive when the queue is full (system with loss) or to adjusting the rate of arrival of messages as a function of queue occupancy rate. In the latter case, the system is said to have a *finite Poisson source* in which the mean number of messages $\lambda(m)$ arriving in the queue in unit time is a function of the number of messages waiting in the queue.

When the queue is part of a network, overload can lead to either routing the arriving messages to other queues in the network or to stopping the arrival of new messages, which are then temporarily stored in the preceding queue. A method often used to take account of the limitations of buffer memories consists of limiting the *total number of messages* circulating in the network to a maximum value M. An application of this technique will be given in the following chapter, in connection with flow control, and a brief indication will be given here of the method of analysing networks of this type which are called *closed exponential networks*.

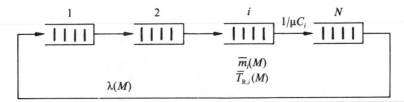

Figure 1.13 Closed exponential network

Consider the simple closed network of Figure 1.13 which consists of N queues and in which a total of M messages is circulating. The mean number of messages waiting in the queue of index i is equal to $\bar{m}_i(M)$ and the mean transfer time of this queue by a message is $\bar{T}_{R,i}(M)$. The mean number of messages circulating per second in the network is equal to $\lambda(M)$. These various parameters clearly depend on the total number M of messages circulating in the network. It will be assumed here that the messages have a mean length of $1/\mu$ and the output line of the queue of index i has a capacity of C_i bit/s. Such a network can be analysed in a relatively simple manner by the **mean value method** [1.12]. This method makes use of the fact that at the time of arrival of a message in the queue, the mean transfer time of the message can be considered to be equal to the service time of this message plus the service time of the messages waiting in the queue at that time

$$\bar{T}_{R,i}(M) = \frac{1}{\mu C_i} + \frac{1}{\mu C_i} \times \left(\begin{array}{c} \text{number of waiting messages} \\ \text{at the time of arrival of the message} \end{array} \right) \quad (1.157)$$

The number of messages waiting in the queue at the time of arrival of the new message is given by the **arrival theorem** which specifies that this number has the same distribution as the steady-state solution of the same closed exponential system containing $M - 1$ messages instead of M. This then leads to a recursive solution of the closed exponential system with

$$\bar{m}_i(0) = 0 \quad (1.158)$$

$$\bar{T}_{R,i}(M) = \frac{1}{\mu C_i} + \frac{\bar{m}_i(M - 1)}{\mu C_i} \quad (1.159)$$

$$\lambda(M) = \frac{M}{\displaystyle\sum_{i=1}^{N} \bar{T}_{R,i}(M)} \quad (1.160)$$

$$\bar{m}_i(M) = \lambda(M)\bar{T}_{R,i}(M), \quad i = 1, \ldots, N \quad (1.161)$$

where (1.160) and (1.161) are obtained by application of Little's formula. This method can be generalized to a more complex network by using a decomposition method which enables it to be reduced to a set of networks of the type represented in Figure 1.13.

1.2 ANALYSIS OF DATA LINKS

1.2.1 Full duplex link operated with a 'send and wait' procedure

The techniques presented in the preceding sections can be used to examine the efficiency of line procedures and to determine the response times of data links. A data link operated with a 'send and wait' procedure and a data link operated with a sliding window protocol will be analysed in turn in order to show that the latter protocol is dramatically better than the former from the standpoint of line throughput [1.16]–[1.21].

Two stations A and B which are interconnected by a two-way simultaneous link will be considered here. The line is operated with a 'send and wait' procedure, which indicates that a station must wait for an acknowledgement of the frame which it has just sent before transmitting the following frame. It will be assumed that the line capacity is C bit/s in both directions and the text frames have an exponentially distributed length with mean lengths of $1/\mu_1$ and $1/\mu_2$ for the A $\rightarrow$ B and B $\rightarrow$ A directions respectively. It will also be assumed that the acknowledgement frames are distinct from the text frames and have a constant length of $1/\nu$ bits.

The maximum useful capacity $C_{\max}$ of the bidirectional link will now be determined. This maximum useful capacity corresponds to text frames which are exchanged when the stations A and B draw on an unlimited amount of frames waiting for transmission. If the text frames have a constant length of $1/\mu$ bits and propagation times are neglected,

$$C_{\max} = \frac{2C}{1 + \mu/\nu} \tag{1.162}$$

The full duplex link can be considered to consist of two one-way links of capacity C each of which carries text during a time $1/\mu C$ and an acknowledgement during $1/\nu C$. If the acknowledgement frames are much shorter than the text frames, then $C_{\max} = 2C$.

When the frames are of variable length, the frame transmitted at a given time by A will generally not have the same length as the frame transmitted at the same time by B. As acknowledgement of the shorter text frame, transmitted in one direction, must wait for transmission until the longer frame has been transmitted, the channel which transmits the shorter frame is idle for part of the time (Figure 1.14). The duplex link operated with a 'send and wait' procedure therefore has a maximum capacity less than that given by (1.162). As a consequence of the synchronization imposed by the procedure, the system represented by two stations and two unidirectional lines can be in one of the four possible states represented in Figure 1.15; this corresponds to transmission of text by the two stations (T, T), transmission of text by a single station, (T, I) and (I, T), and transmission of acknowledgements by the two stations (ACK, ACK). When the length of the frames is exponentially distributed, the transmission time of a frame sent by A is between t and $t + \mathrm{d}t$ with a probability $f_1(t)$ given by

$$f_1(t) = \mu_1 C \, e^{-\mu_1 C t} \tag{1.163}$$

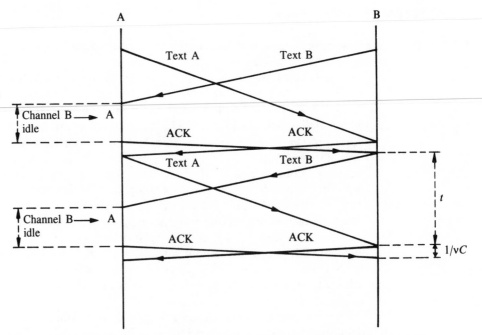

Figure 1.14 Full duplex link operated with a 'send and wait' procedure

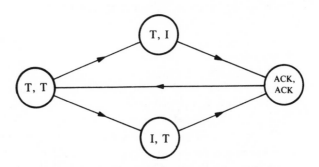

Figure 1.15 Possible states of a duplex link

with a similar equation for the probability density function $f_2(t)$ which corresponds to messages transmitted from B to A. The links are in text transmission phases (T, T), (T, I) and (I, T) for a time t if the duration of transmission in one direction is equal to t and the duration of transmission in the other direction is less than t. The latter condition occurs with a probability $P_1(\tau < t)$ given by

$$P_1(\tau < t) = \int_0^t \mu_1 C e^{-\mu_1 C \tau} \, d\tau = 1 - e^{-\mu_1 C t} \tag{1.164}$$

with a similar relation for $P_2(\tau < t)$. The text transmission phase thus has a duration between t and $t + dt$ with a probability $f(t)$ given by

$$f(t) = f_1(t)P_2(\tau < t) + f_2(t)P_1(\tau < t)$$ (1.165)

or, using (1.163) and (1.164),

$$f(t) = \mu_1 Ce^{-\mu_1 Ct}(1 - e^{-\mu_2 Ct}) + \mu_2 Ce^{-\mu_2 Ct}(1 - e^{-\mu_1 Ct})$$ (1.166)

The last expression is a sum of exponential laws. The mean duration $\bar{t}$ of the text transmission phase can thus be determined as in § 1.1.2 which gives

$$\bar{t} = \frac{1}{\mu_1 C} + \frac{1}{\mu_2 C} - \frac{1}{\mu_1 C + \mu_2 C}$$ (1.167)

The duration of transmission of acknowledgements is $1/vC$ and the mean duration of transmission of text frames in each direction is $1/\mu_1 C$ and $1/\mu_2 C$. The maximum link capacity is thus given by

$$C_{max} = \frac{C\left(\dfrac{1}{\mu_1 C} + \dfrac{1}{\mu_2 C}\right)}{\dfrac{1}{vC} + \dfrac{1}{\mu_1 C} + \dfrac{1}{\mu_2 C} - \dfrac{1}{\mu_1 C + \mu_2 C}}$$ (1.168)

If the mean length of the text frames is the same in the two directions ($1/\mu_1 = 1/\mu_2 = 1/\mu$) and the length of acknowledgement frames is very much less than the mean length of text frames, Equation (1.168) reduces to

$$C_{max} = 4C/3$$ (1.169)

It can be seen that a full duplex line (of capacity $2C$) is at best used to only 67 per cent of its maximum capacity when it is operated as a two-way simultaneous link managed by a 'send and wait' procedure. In the general case of frames of different lengths, the maximum throughput efficiency E of the duplex line, expressed as a percentage, is obtained directly from (1.168) as

$$E = \frac{\dfrac{50}{\mu_1}\left(1 + \dfrac{\mu_1}{\mu_2}\right)}{\dfrac{1}{v} + \dfrac{1}{\mu_1}\left(1 + \dfrac{\mu_1}{\mu_2} - \dfrac{\mu_1/\mu_2}{1 + \mu_1/\mu_2}\right)}$$ (1.170)

By way of example, Figure 1.16 shows the throughput efficiency of the duplex line for acknowledgement frames having a length of six bytes. It can be seen that the efficiency is maximum when the mean length of text frames is the same in both directions; this is not surprising since one of the channels remains idle while waiting for acknowledgement when the traffic is not symmetrical. In practice, when the lines are heavily loaded, the analysis must take propagation delays into account and this can be achieved by a relatively simple modification to the previous equations.

1.2.2 Full duplex link operated with a sliding window protocol

It was seen in Chapter 4 (Vol. I) that the utilization efficiency of a line can be significantly improved by replacing the 'send and wait' procedure with a sliding

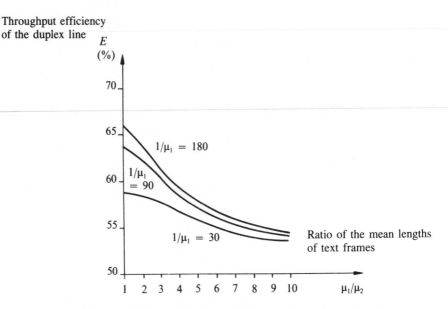

Figure 1.16 Maximum throughput efficiency of a duplex line operated as a two-way link with a 'send and wait' procedure. Length of acknowledgement frames: $1/\nu = 6$ bytes

window protocol. With the latter approach, stations try to avoid the blocking which is inherent in the 'send and wait' procedure by anticipating the expected acknowledgements. In this way, with the HDLC procedure, stations can send up to seven (or 127) consecutive frames before receiving acknowledgement of some or all of these frames. The degree of anticipation, that is the number of frames which can be sent before acknowledgement, is the **window W** of the procedure. It can easily be seen that blocking can be almost completely eliminated if W is chosen to be sufficiently large. In practice, it is not possible to set W to too high a value since this would lead to an excessive size of the transmitter buffers and an excessive number of retransmissions in the case of errors on the line. It will be shown here, using a simplified analysis [1.22], that the maximum values of W selected for the HDLC procedure, which are 7 and 127 for short and long haul links respectively, allow near optimum line utilization.

As in the above section, a two way simultaneous link will be considered. It will also be assumed that acknowledgements of one channel are inserted into information frames of the other channel if there are any, otherwise they are sent in separate supervisory frames (HDLC procedure). One of the two unidirectional channels will be analysed by assuming that the other is never blocked; it will be assumed, as above, that the transmitter of the channel analysed has an adequate store of information frames to be transmitted so that it never has to wait for the arrival of such a frame before transmission. The case where the information frames have a constant length of l bits will be considered first; let $\bar{T}_R$ be the mean time between the start of transmission of an information frame and receipt of the corresponding acknowledgement. Blocking will not be possible if W is sufficiently large for the transmitter to continue sending information frames up to the arrival of an acknowledgement. The non-blocking condition is thus given by

$$\frac{Wl}{C} > \bar{T}_R \qquad (1.171)$$

where C is the capacity of the channel expressed in bit/s. To evaluate $\bar{T}_R$, it will first be noted (Figure 1.17) that the time between the start of transmission by A of an information frame I_A, and the end of its receipt by B is equal to the transmission time l/C of I_A plus the propagation delay T_p (T_p also corresponds to the processing time within receiver B). Once the frame I_A is received by B, a time τ will, in general, elapse before transmission of the frame from B to A is completed. At the end of this frame, B transmits the acknowledgement, either by inserting it in an information frame I_B or by sending a supervisory frame of length l_a bits. This corresponds to a transmission time T_a. Finally, the acknowledgement is received by A after a propagation time T_p. Hence:

$$\bar{T}_R = \frac{l}{C} + 2T_p + \tau + T_a \qquad (1.172)$$

If the channel on which the acknowledgements are transmitted was used permanently, the waiting time τ would be equal on average to half the transmission time of an information frame, that is $l/2C$. It is accepted here that the return channel transmits information frames with a probability p. Hence

$$\tau = pl/2C \qquad (1.173)$$

$$T_a = \frac{pl}{C} + (1 - p)\frac{l_a}{C} \qquad (1.174)$$

from which, using (1.171) and (1.172),

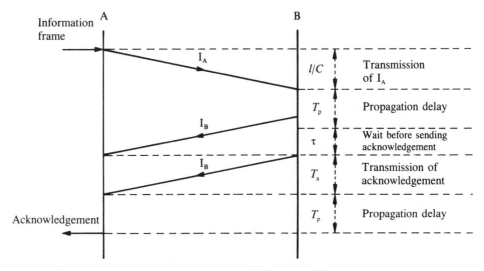

Figure 1.17 Transmission time $\bar{T}_R$

$$W > 1 + \frac{3p}{2} + (1 - p)\frac{l_a}{l} + \frac{2CT_p}{l} \qquad (1.175)$$

When the return channel is used to its full capacity, $p = 1$, and W becomes

$$W > 2.5 + \frac{2CT_p}{l} \qquad (1.176)$$

When the frames are of variable length, analysis of the link is much more difficult and an approach based on mean values can give only a crude approximation. Notice, however, that in this case the waiting time τ can be evaluated by considering that the ends of frame transmission on the return channel can be regarded as part of a renewal process in which the delays between events are independent random variables and are identically distributed. This leads to definition of τ by

$$\tau = \frac{p \, \overline{l^2}}{2\bar{l}C} \qquad (1.177)$$

where $\bar{l}$ is the mean length of the information frames. It is also necessary to introduce a correcting factor of the form $\overline{l^2}/\bar{l}^2$ to take account of the different random variables which play a part in the process. This leads to a value of W defined by

$$W > \left[1 + p + p\frac{\overline{l^2}}{2\bar{l}^2} + \frac{2CT_p}{l} + (1 - p)\frac{l_a}{l} \right]\frac{\overline{l^2}}{\bar{l}^2} \qquad (1.178)$$

When the return channel is used to full capacity, $p = 1$ and W becomes

$$W > \left(2 + \frac{\overline{l^2}}{2\bar{l}^2} + \frac{2CT_p}{l} \right)\frac{\overline{l^2}}{\bar{l}^2} \qquad (1.179)$$

If the distribution of frame lengths is exponential (§ 1.1.2),

$$\frac{\overline{l^2}}{\bar{l}^2} = 2 \qquad (1.180)$$

and W becomes, for $p = 1$,

$$W > 2\left(3 + 2\frac{CT_p}{l} \right) \qquad (1.181)$$

By way of example, Table 1.3 gives the minimum window size required to prevent blocking in the case of a medium range link (500 km, 4800 bit/s) and also for a satellite link (64 kbit/s). In the two cases, it has been assumed that the return channel is fully loaded and the values of W have been found from (1.176) and (1.181). It can be seen that, except for short frames transmitted on long haul lines, the values $W = 7$ and $W = 127$ are sufficient to prevent blocking. These results are confirmed by more thorough analysis of the operation of the procedure [1.18].

Table 1.3 Minimum size of the window W to prevent blocking.

		Length of information frames (bits)					
		64	128	256	512	1024	
500 km 4800 bit/s	Fixed Frames	2.9	2.7	2.6	2.5	2.5	W
$T_p = 2.5$ ms	Exponential frames	6.8	6.4	6.2	6.1	6	W
Satellite 64 kbit/s	Fixed frames	542	272	137	70	36	W
$T_p = 270$ ms	Exponential frames	1086	546	276	141	73	W

1.3 CONCENTRATORS AND MULTIPLEXERS

1.3.1 Analysis of a concentrator

Packet switching networks consist of nodes which are interconnected by transmission lines. The packets which travel on each line are distinct from each other and can have very different destinations. Every node must therefore route every incoming packet to the output line which will direct it towards its destination (Figure 1.18). The network can thus be considered to consist of multiplexed lines connected to statistical concentrators located at the nodes. Because of the random arrival of packets at the node, it is necessary to provide a buffer on each output line in order to avoid any loss of packets when these arrive simultaneously. The model of the concentrator thus contains as many queues as there are output lines,

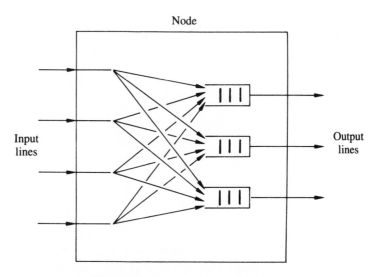

Figure 1.18 Concentrator–multiplexer

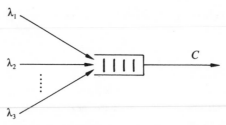

Figure 1.19 Queue with multiple entry

and each queue is fed from the N node input lines which provide it with messages destined for its output line (Figure 1.19).

The case will be considered here of a queue of this type with a geometric distribution of message lengths and Poisson arrivals of mean value $\lambda_1, \lambda_2, \ldots, \lambda_N$ messages per second on the various input lines. Such a queue can thus be considered as a generalization of the queue which was examined in § 1.1.8. Under these conditions, if C is the capacity of the output line expressed as a number of words per second (each message contains l words, where l is a random variable defined by the geometric distribution) and n_i is the number of words entering the queue by line i during the interval $\Delta t = 1/C$, the total number n of words entering the queue during the interval $\Delta t = 1/C$ is given by

$$n = \sum_{i=1}^{N} n_i \tag{1.182}$$

As in § 1.1.8, it is easy to show that the mean number $\bar{m}$ of words waiting in the queue is given by Equation (1.91) which is recalled here

$$\bar{m} = \frac{\bar{n}}{2} + \frac{V(n)}{2(1 - \bar{n})} \tag{1.183}$$

with, in this case

$$\bar{n} = \sum_{i=1}^{N} \bar{n}_i \tag{1.184}$$

and

$$V(n) = \sum_{i=1}^{N} V(n_i) \tag{1.185}$$

Equations (1.184) and (1.185) with (1.102) and (1.105) give

$$\bar{n} = \sum_{i=1}^{N} \frac{\lambda_i \bar{l}_i}{C} \tag{1.186}$$

$$V(n) = \sum_{i=1}^{N} \frac{\lambda_i \bar{l}^{2}}{C} \left(2 - \frac{1}{\bar{l}_i}\right) \tag{1.187}$$

where $\bar{l}_i$ is the mean number of words per message on input line i. By evaluating (1.183) with (1.186) and (1.187), the mean number of words waiting in the queue is obtained

$$\bar{m} = \frac{\sum_{i=1}^{N} \lambda_i \bar{l}_i}{2C} + \frac{\sum_{i=1}^{N} \lambda_i \bar{l}_i (2\bar{l}_i - 1)}{2\left(C - \sum_{i=1}^{N} \lambda_i \bar{l}_i\right)} \qquad (1.188)$$

The mean number $\lambda\bar{l}$ of words entering the queue is equal to the sum of the mean arrivals on each line, hence

$$\lambda\bar{l} = \sum_{i=1}^{N} \lambda_i \bar{l}_i \qquad (1.189)$$

Under these conditions, the mean transfer time $\bar{T}_R$ of the queue by a message is deduced from $\bar{m}$ using Little's formula:

$$\bar{T}_R = \bar{m}/(\lambda\bar{l}) \qquad (1.190)$$

$$\bar{T}_R = \frac{1}{2C} + \frac{\sum_{i=1}^{N} \lambda_i \bar{l}_i (2\bar{l}_i - 1)}{2\left(C - \sum_{i=1}^{N} \lambda_i \bar{l}_i\right) \sum_{i=1}^{N} \lambda_i \bar{l}_i} \qquad (1.191)$$

When the messages have the same mean length $\bar{l}$ on each input line, the mean number of waiting words and the transfer time are the same as for a queue with a single input line, on condition that the arrival rate λ of messages is considered to be equal to the sum of the arrivals of messages on the various lines

$$\lambda = \sum_{i=1}^{N} \lambda_i \qquad (1.192)$$

$$\bar{m} = \frac{\lambda\bar{l}}{2C} + \frac{\bar{l}\lambda(2\bar{l} - 1)}{2(C - \lambda\bar{l})} \qquad (1.193)$$

$$\bar{T}_R = \frac{1}{2C} + \frac{2\bar{l} - 1}{2(C - \lambda\bar{l})} \qquad (1.194)$$

1.3.2 Overflow probability

So far it has been assumed that the queue buffers have an infinite size. In practice, buffers always have a finite size and the designer's problem is to determine the buffer size as a function of the expected traffic and to calculate the delay introduced by the concentrator. A first approach to this problem consists of determining a buffer size which is sufficiently large for the *overflow probability* of the buffer to be very small. In this case, the mean transfer time of the queues is very close to that corresponding to an infinite memory, and it is thus given by Equations (1.191) and (1.194). Computation of the overflow probability is rather difficult in the case of a geometric distribution of message length [1.23] and consideration here will be restricted to the case of M/M/1 queues which correspond to an exponential distribution of message length. It has already been shown that the latter distribution gives results very close to the geometric distribution for messages having a long mean length and it is required to obtain only an estimate for the buffer size in order to limit the overflow probability to less than a given value.

It was shown in § 1.1.4 that the probability that the number of messages waiting in the queue exceeds M_1 is equal to ρ^{M_1+1}, with $\rho = \lambda/\mu C$. As the mean number of words is approximately equal to the product of the mean number of messages and the mean message length $\bar{l}$, the probability $P(m > M)$ that the number of words waiting in the queue exceeds M is given by

$$P(m > M) = \rho^{1+M/l} \tag{1.195}$$

with

$$\rho = \lambda \bar{l}/C \tag{1.196}$$

where C is the capacity of the output line expressed as a number of words per second. The concentrator can thus be designed by using (1.195) to select a buffer size M such that $P(m > M)$ is less than a given value, for example 10^{-3}. The mean number of waiting words and the mean transfer time are then calculated by using Equations (1.193) and (1.194).

1.3.3 Concentrator with finite memory

The concentrator can be examined in a more realistic manner by proceeding as for an infinite queue, but limiting the number of waiting words to the maximum buffer size M. As in the above section, it is possible to obtain simply an estimate of the overflow probability p_B by assuming that the messages have an exponentially distributed length. In this case, the queue is of the M/M/1 type and it was shown in § 1.1.4 that, for a buffer with a maximum capacity of M_1 messages, the blocking probability p_B was given by Equation (1.43) which is recalled here

$$p_B = \frac{(1 - \rho)\rho^{M_1}}{1 - \rho^{M_1+1}} \tag{1.197}$$

As $M_1 \bar{l} = M$ approximately, the overflow probability is given here in terms of the buffer size expressed as a function of the maximum number of words M, hence

$$p_B = \frac{(1 - \rho)\rho^{M/l}}{1 - \rho^{1+M/l}} \tag{1.198}$$

and, as above

$$\rho = \lambda \bar{l}/C \tag{1.199}$$

It is possible to define a more exact model of the concentrator by considering messages of discrete length [1.24], [1.25] and it will be shown here how the overflow probability can be calculated in the case of a geometric distribution of message lengths. In § 1.1.8, a queue was examined in which message arrivals are Poisson and the messages have a geometrically distributed length with a mean length of $\bar{l}$. It has been seen that, in this case, the probability distribution q_i for

i words to enter during the time interval $1/C$ is defined by the moment-generating function $H(z)$ given by (1.99) which is written here as

$$H(z) = e^{-\lambda/C}\, e^{\lambda z/\{C[\bar{l}-(\bar{l}-1)z]\}} \tag{1.200}$$

since $\bar{l} = 1/p$ for a geometric distribution. By definition of the moment-generating function, we have

$$H(z) = \sum_{i=0}^{\infty} q_i z^i \tag{1.201}$$

It is thus possible to compute the various values of q_i for $i = 0, 1, 2, \ldots,$ by successive differentiation of (1.200) since

$$q_i = \frac{1}{n!} \left. \frac{d^n H(z)}{dz^n} \right|_{z=0} \tag{1.202}$$

In practice, it is in general more convenient to calculate the q_i by the classical method of the inverse Fourier transform [1.24]. Once the q_i are known, it is possible to calculate, for all values of i, where $i = 0, 1, \ldots,$ the probability p_i of having i words waiting in the queue in the steady state. As the capacity of the output line is C words per second, the queue will be emptied of one word after a time interval $1/C$, on condition that it is not empty at the start of this interval. During this same time interval $1/C$, i words arrive with a probability q_i, except when the queue is already full at the start of the interval, in which case no new word can enter the queue during the interval $1/C$. The probabilities p_i are thus defined by

$$p_0 = q_0 p_1 + q_0 p_0$$

$$p_1 = q_0 p_2 + q_1 p_1 + q_1 p_0$$

$$\vdots$$

$$p_u = \sum_{i=1}^{u+1} q_{u+1-i} p_i + q_u p_0$$

$$\vdots$$

$$p_M = p_M(1 - q_0) + p_{M-1}(1 - q_0 - q_1)$$

$$+ (p_1 + p_0)\left(1 - \sum_{i=0}^{M-1} q_i\right)$$

$$p_{u>M} = 0 \tag{1.203}$$

with clearly

$$\sum_{i=0}^{M} p_i = 1 \tag{1.204}$$

The various probabilities p_i can be computed recursively as a function of p_0 using (1.203), and p_0 can itself be calculated from (1.204). The overflow probability p_B can then be determined by noting that the mean number of words which arrive in the queue during the interval $1/C$ is $\lambda \bar{l}/C$. However, as words cannot enter the queue when it is full, only $(1 - p_{\mathrm{B}})\lambda \bar{l}/C$ words enter the queue on average during an interval $1/C$. During this same interval, the output line extracts a word from the queue, except when the latter is empty. Hence, in the steady state

$$(1 - p_B)\frac{\lambda \bar{l}}{C} = 1 - p_0 \tag{1.205}$$

$$p_B = 1 - \frac{(1 - p_0)C}{\lambda \bar{l}} \tag{1.206}$$

When the overflow probability is sufficiently small, for example when it is less than 10^{-3}, the behaviour of the queue with Poisson message arrivals and geometric distribution of message length is close to that of the corresponding queue with infinite memory, and the mean transfer time $\bar{T}_{\mathrm{R}}$ of the queue by a message can be evaluated using Equation (1.194) which is recalled here

$$\bar{T}_{\mathrm{R}} = \frac{1}{2C} + \frac{2\bar{l} - 1}{2(C - \lambda \bar{l})} \tag{1.207}$$

1.3.4 Example of the analysis and design of a concentrator

In order to illustrate the technique used to design a concentrator and analyse its performance, the example will be considered of a simple system consisting of 50 terminals connected to a computer by way of a concentrator (Figure 1.20). It will be assumed that each terminal sends a message of mean length 10 bytes to the computer every 3 seconds on average. The system is intended to be operated in interactive mode, so the computer replies to each message which is addressed to it with a response whose mean length is 20 bytes. The concentrator is connected to the computer by a two-way simultaneous link with a capacity of 4800 bit/s in each direction. To take account of additional traffic which must be provided to

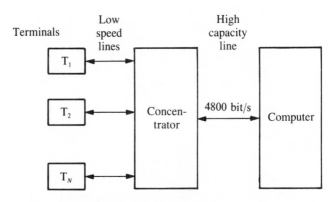

Figure 1.20 Example of a concentrator

support the service functions of the system (for example, the terminal address, supervisory and frame delimiting information), it will be assumed that the service traffic leads to a 20 per cent increase in the mean message length. The messages must therefore be considered to have a mean length of 12 bytes in the terminal–computer direction and 24 bytes in the computer–terminal direction.

With these assumptions, the capacity C of the line connecting the computer to the concentrator is $4800/8 = 600$ bytes per second. Consider, firstly, the operation of the concentrator in the terminal–computer direction. The total traffic intensity is given by

$$\rho = \frac{\lambda \bar{l} N}{C} = \frac{12 \cdot 50}{3 \cdot 600} = \frac{1}{3} \tag{1.208}$$

It will be assumed henceforth that the message arrivals are Poisson. To determine the size M to give to the buffer so that the overflow probability is less than a given value, a first approach is to consider the case of a queue of infinite size with an exponential distribution of message lengths. Then, from (1.195),

$$M = \bar{l} \frac{(\log p_B - \log \rho)}{\log \rho} \tag{1.209}$$

For $p_B = 10^{-3}$, the buffer size, expressed as a number of bytes, is given by

$$M = 12 \frac{(3 - 0.477)}{0.477} \simeq 64 \tag{1.210}$$

Similarly, it is found that M must be equal to 139 bytes for the overflow probability to be less than 10^{-6}. A more precise value of the size to be given to the buffer can be found by considering a queue of finite size directly, still with an exponential message length distribution. M is then calculated from (1.198):

$$M = \bar{l} \frac{(\log p_B - \log (1 - \rho + \rho p_B))}{\log \rho} \tag{1.211}$$

which gives $M = 71$ and $M = 147$ for $p_B = 10^{-3}$ and $p_B = 10^{-6}$ respectively. The buffer size which must be provided for the queue can be calculated more precisely by introducing a geometric distribution for the message lengths, but it can already be seen from the previous calculation that the buffer size must be of the order of 150 to 200 bytes for the probability of loss of a message due to buffer overflow to be negligible. In this case, the mean transfer time $\bar{T}_R$ can be evaluated using (1.207) which gives

$$\bar{T}_R = \frac{1}{2 \cdot 600} + \frac{2 \cdot 12 - 1}{2 \left(600 - \frac{50 \cdot 12}{3} \right)} = 0.03 \text{ s} \tag{1.212}$$

Now consider the operation of the concentrator in the computer–terminal direction. The calculations are the same as previously, but with a mean message length $\bar{l}$ of 24 bytes instead of 12. The total traffic intensity ρ is thus equal to $\frac{2}{3}$. By proceeding in the same manner as above, $M = 385$ and $M = 794$ are obtained for overflow probabilities of 10^{-3} and 10^{-6} and an infinite queue. Similarly,

Equation (1.211) gives $M = 344$ and $M = 753$ for $p_B = 10^{-3}$ and $p_B = 10^{-6}$ respectively in the case of a finite M/M/1 model. It can thus be seen that the buffer size must be of the order of 700 bytes for the probability of loss to be negligible. In this case, the mean transfer time of the queue by a message is given by (1.207),

$$\bar{T}_R = \frac{1}{2 \cdot 600} + \frac{2 \cdot 24 - 1}{2 \left(600 - \frac{50 \cdot 24}{3} \right)} = 0.12 \text{ s} \tag{1.213}$$

1.4 OPTIMIZATION OF NETWORK TOPOLOGY

1.4.1 General principles

On account of the high cost of communication, the topology of teleprocessing networks must be examined very carefully; even if successful optimization of the network enables costs to be reduced by only a small factor, the amount saved can be considerable. The problem of optimizing a teleprocessing network can be formulated as follows: given the data processing equipment to be interconnected, its location and the traffic between stations, together with the cost of the links and the various equipment on the network, minimize the cost of the network. This optimization must be performed while taking account of a number of constraints of which the principal ones are generally to ensure network availability and to guarantee that the network packet transfer time does not exceed a given limit. To carry out this optimization, the designer can change the topology; that is, he can decide on the switching and multiplexing equipment to be used together with the lines to be established between subscriber's equipment and the network equipment. The designer can also change the capacity of the network lines and adopt a policy of his choice in relation to the routing of packets.

The problem thus defined cannot be resolved by an exhaustive search, since the number of possible combinations becomes enormous as soon as the network contains more than a few nodes. Furthermore, the complexity of the problem is such that there is no analysis method which leads directly to an optimum network. The solution which is generally adopted, and will be described here, consists of dividing the problem into a series of sub-problems which are treated separately, either by analytic methods or heuristic methods, and involve a large amount of computer calculation.

The first simplification which is generally used in network design is to divide the latter into two parts (Figure 1.21), the **transit network** and the **local network**, and to treat the design of these two parts separately. The transit network is the heart of the teleprocessing network and consists of high-capacity lines which link the nodes, each of which contains a packet switching multiplexer. The local network connects subscriber's stations to the nodes of the transit network either directly or by way of concentrators or multipoint lines.

The design of the transit network will be considered first by assuming that its topology is known and attempting to determine the line capacities which minimize

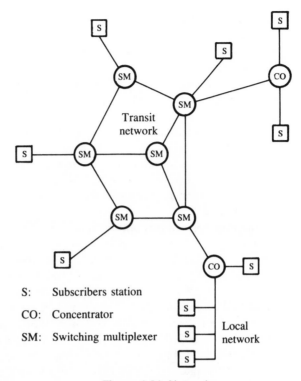

Figure 1.21 Network

the cost of the network. Once this problem of capacity allocation is solved, it will be shown how the topology of the network can be modified to approach a minimum cost solution. The problem of local network design will then be tackled by attempting to optimize the multipoint lines and showing how to determine the number and geographical position of the concentrators.

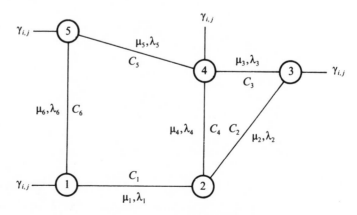

Figure 1.22 Transit network

1.4.2 Capacity allocation in a transit network

The transit network is defined here by its topology (Figure 1.22) and by the external traffic $\gamma_{i,j}$ which it must transfer between each pair of input–output nodes i, j. For simplicity, it will be assumed that the traffic is symmetrical, that is the same in the two directions and $\gamma_{i,j} = \gamma_{j,i}$. In practice, this assumption is not restrictive since the method which will be used extends very simply to the case of asymmetric traffic by considering transmission in each direction separately for each link. $\gamma_{i,j}$ is expressed as a number of packets per second and represents a particular traffic for which the mean length of packets is given. The capacity C of the various lines will now be determined, that is the capacity which must be assigned to them to minimize the cost of the network for a given mean packet transfer time.

The first stage of this optimization is to determine the mean arrival of messages λ_u together with the mean message length $1/\mu_u$ on each of the M lines which interconnect the N network nodes. To do this it is necessary to know the **routing policy**, that is the paths followed by the various packets. The routing policy can be static or dynamic and it normally results from an optimization that will be described in Chapter 2. At the present stage, it will be assumed that the routing of messages is fixed and known. Under these conditions, the data concerning the traffic to be carried by the network can be summarized in a traffic matrix of the type given in Table 1.4. Here each cell indicates the traffic $\gamma_{i,j}$ and the path followed for packets which enter the network by node i, corresponding to the row of the cell, and leave the network by node j, corresponding to the column of the cell. If the packets have different mean lengths according to the particular traffic $\gamma_{i,j}$ to which they belong, the cells of the traffic matrix must also contain the corresponding mean lengths. Hence, with the example of Table 1.4 where all packets have the same mean length, the traffic which enters the network by node

Table 1.4 Traffic matrix of the network of Figure 1.22.

		Destination				
		1 Geneva	2 Lausanne	3 Zurich	4 Bern	5 Basle
Source	1 Geneva		1–2 10 p/s	1–2–3 8 p/s	1–2–4 12 p/s	1–5 5 p/s
	2 Lausanne	2–1 10 p/s		2–3 15 p/s	2–4 6 p/s	2–4–5 2 p/s
	3 Zurich	3–2–1 8 p/s	3–2 15 p/s		3–4 4 p/s	3–4–5 12 p/s
	4 Bern	4–2–1 12 p/s	4–2 6 p/s	4–3 4 p/s		4–5 7 p/s
	5 Basle	5–1 5 p/s	5–4–2 2 p/s	5–4–3 12 p/s	5–4 7 p/s	

1 (Geneva) with a destination of node 3 (Zurich) must pass through node 2 (Lausanne) and it flows with a mean capacity of 8 packets per second.

It will now be assumed that arrivals of packets are distributed according to the Poisson law and the lengths of the packets are exponentially distributed. These assumptions may seem arbitrary, but they can be justified theoretically by *Kleinrock's independence hypothesis* which assumes that packets lose their identity on arrival at a node and the latter reconstructs the packets with an exponentially distributed length. These assumptions are well justified by experience.

Under these conditions, if all the packets circulating in the network have the same mean length, the mean number λ_u of packets which arrive per second on each of the lines u, with $u = 1, \ldots, M$, is determined very simply by taking the sum of all the traffic $\gamma_{i,j}$ which passes by this line. Thus, with the example of Table 1.4, the mean traffic λ_4 on line 4 (Lausanne–Bern) is equal to the sum of the external traffic, $\gamma_{1,4} = 12$ p/s, $\gamma_{2,4} = 6$ p/s and $\gamma_{2,5} = 2$ p/s, which correspond to packets 1–4 (Geneva–Bern), 2–4 (Lausanne–Bern) and 2–5 (Lausanne–Basle) respectively. The mean values of the traffic on the various lines of the network of Figure 1.22 are given by way of example in Table 1.5. When all the packets of the network have the same mean length, then

$$\lambda_u = \sum_{\substack{\text{line} \\ u}} \gamma_{i,j} \tag{1.214}$$

When the packets have different mean values $1/\mu_{i,j}$ according to the pair of nodes i, j which serve them, the mean traffic can again be calculated using (1.214) but the mean length $1/\mu_u$ of the packets on each line of index u must be evaluated by weighting each mean length by the corresponding traffic using

$$\frac{1}{\mu_u} = \frac{1}{\displaystyle\sum_{\substack{\text{line} \\ u}} \gamma_{i,j}} \sum_{\substack{\text{line} \\ u}} \frac{\gamma_{i,j}}{\mu_{i,j}} = \frac{1}{\lambda_u} \sum_{\substack{\text{line} \\ u}} \frac{\lambda_{i,j}}{\mu_{i,j}} \tag{1.215}$$

Table 1.5 Optimum capacity allocation of the network of Figure 1.22. Columns 2 and 3: results for mean transfer time $\overline{T} = 100$ ms. Columns 4 and 5: results for $K = \infty$ and $T_{(\infty)} = 60$ ms.

Line	λ_u p/s	$K = 1$ $T_{(1)} = 100$ ms $C_u\vert_{\text{opt}}$ bit/s	$\overline{T}_u\vert_{\text{opt}}$ ms	$K = \infty$ $T_{(\infty)} = 60$ ms $C_u\vert_{\text{opt}}$ bit/s	$\overline{T}_u\vert_{\text{opt}}$ ms
1. Geneva–Lausanne	30	12105	59	11946	60
2. Lausanne–Zurich	23	9762	66	10155	60
3. Zurich–Bern	16	7327	79	8363	60
4. Lausanne–Bern	20	8733	71	9387	60
5. Bern–Basle	21	9078	69	9643	60
6. Geneva–Basle	5	3086	142	5547	60

If line 4 (Lausanne–Bern) of the preceding example is reconsidered, assuming this time that the mean lengths of the packets depend on the source–destination pair and are 200 bits, 300 bits and 500 bits for packets 1–4, 2–4 and 2–5 respectively, the mean traffic is still 20 p/s but the mean length $1/\mu_4$ of the messages must be calculated as

$$\frac{1}{\mu_4} = \frac{1}{20}\,(200\cdot 12 + 300\cdot 6 + 500\cdot 2) = 260$$

After these various operations, the mean traffic λ_u is known for each of the M lines of the network together with the mean packet length $1/\mu_u$ and it is now necessary to determine the capacity C_u which must be provided on each line to minimize the cost of the network while ensuring a mean transfer time $\bar{T}$ less than a given limit.

As the queues are of the M/M/1 type, the mean transfer time $\bar{T}_u$ on the line of index u is given by (1.36) as

$$\bar{T}_u = \frac{1}{\mu_u C_u - \lambda_u} \tag{1.216}$$

where $1/\mu_u$ is expressed as a number of bits and C_u is expressed as a number of bits per second. The total mean traffic γ which enters the network is defined by

$$\gamma = \sum_{i=1}^{N}\sum_{j=1}^{N} \gamma_{i,j} \tag{1.217}$$

Similarly, the total mean traffic λ on the network lines is given by

$$\lambda = \sum_{u=1}^{M} \lambda_u \tag{1.218}$$

λ and γ in general have different values since the packets most often have to pass through several nodes to reach their destination. If $n_{i,j}$ is the number of nodes which the packets corresponding to the traffic $\gamma_{i,j}$ must pass through, the mean traffic λ on the lines of the network is given by

$$\lambda = \sum_{i=1}^{N}\sum_{j=1}^{N} n_{i,j}\,\gamma_{i,j} \tag{1.219}$$

λ is thus equal to γ only if $n_{i,j} = 1$ for all i and j, that is only when there is a direct link between each source–destination pair. The mean number $\bar{n}$ of nodes crossed by the packets in the whole network is clearly equal to the weighted sum of the number of nodes crossed by the various source–destination pairs. Hence

$$\bar{n} = \frac{\displaystyle\sum_{i=1}^{N}\sum_{j=1}^{N} n_{i,j}\gamma_{i,j}}{\displaystyle\sum_{i=1}^{N}\sum_{j=1}^{N} \gamma_{i,j}} = \frac{\lambda}{\gamma} \tag{1.220}$$

The mean transfer time $\bar{T}$ of the network by a packet is thus equal to the product of the number of lines passed through and the mean transfer time of the line. As the number of lines passed through is the same as the number of nodes crossed,

$$\bar{T} = \frac{\bar{n}}{\lambda} \sum_{u=1}^{M} \lambda_u \bar{T}_u = \frac{1}{\gamma} \sum_{u=1}^{M} \frac{\lambda_u}{\mu_u C_u - \lambda_u} \qquad (1.221)$$

It will now be assumed that the cost of the lines is linearly related to their capacity. This could, for example, correspond to the case where the PTT lease their lines on the basis of a fixed monthly connection charge b_u plus a monthly charge a_u which is proportional to capacity. On this assumption, the total cost F of the network lines is given by

$$F = \sum_{u=1}^{M} (a_u C_u + b_u) \qquad (1.222)$$

The values of C_u will be determined to minimize the total cost F for a given value of the mean transfer time $\bar{T}$. This problem can be solved very simply by the method of **Lagrangian multipliers** by noting that, since the difference between the first and second terms of (1.221) is zero, Equation (1.222) is still valid if it is written as follows:

$$F = \sum_{u=1}^{M} (a_u C_u + b_u) - \alpha \left(\bar{T} - \frac{1}{\gamma} \sum_{u=1}^{M} \frac{\lambda_u}{\mu_u C_u - \lambda_u} \right) \qquad (1.223)$$

where α is the Lagrangian multiplier.

The minimum of F is thus obtained by finding values of C_u which make the partial derivatives of F with respect to C_u equal to zero

$$\frac{\partial F}{\partial C_u} = a_u - \frac{\alpha \lambda_u \mu_u}{\gamma(\mu_u C_u - \lambda_u)^2} = 0 \qquad (1.224)$$

or

$$\frac{1}{\mu_u C_u - \lambda_u} = \left(\frac{\gamma a_u}{\alpha \lambda_u \mu_u} \right)^{1/2} \qquad (1.225)$$

By mutliplying the two terms of (1.225) by λ_u/γ and summing over all lines, the following is obtained

$$\frac{1}{\gamma} \sum_{u=1}^{M} \frac{\lambda_u}{\mu_u C_u - \lambda_u} = \bar{T} = \frac{1}{\gamma} \sum_{u=1}^{M} \left(\frac{\gamma \lambda_u a_u}{\alpha \mu_u} \right)^{1/2} \qquad (1.226)$$

from which

$$\alpha^{1/2} = \frac{1}{\gamma \bar{T}} \sum_{u=1}^{M} \left(\frac{\gamma \lambda_u a_u}{\mu_u} \right)^{1/2} \qquad (1.227)$$

and, using (1.225),

$$C_u \bigg|_{\text{optimum}} = \frac{\lambda_u}{\mu_u} + \frac{1}{\gamma \bar{T}} \left(\frac{\lambda_u}{\mu_u a_u} \right)^{1/2} \sum_{u=1}^{M} \left(\frac{\lambda_u a_u}{\mu_u} \right)^{1/2} \qquad (1.228)$$

By substituting the values of C_u defined by (1.228) into (1.222), the minimum cost $F_{\min}$ is obtained

$$F_{\min} = \sum_{u=1}^{M} \left(\frac{a_u \lambda_u}{\mu_u} + b_u \right) + \frac{1}{\gamma \bar{T}} \left[\sum_{u=1}^{M} \left(\frac{\lambda_u a_u}{\mu_u} \right)^{1/2} \right]^2 \qquad (1.229)$$

By way of example, the network defined by Figure 1.22 and Table 1.4 will be considered again and the capacity allocation which leads to minimum cost will be determined by assuming that $b_u = 0$ and $a_u = 1$. In this case, the total cost F is the same as the total capacity of the network C

$$C = \sum_{u=1}^{M} C_u \qquad (1.230)$$

It will also be assumed that the mean length of the packets is 256 bits throughout the network. By adding the input traffic of the traffic matrix (Table 1.4), the total external traffic is obtained, with $\gamma = 81$. Similarly, the total internal traffic of the network can be calculated by summing the values of λ_u already obtained above (Table 1.5). This gives $\lambda = 115$ and permits calculation of the mean number of lines passed through by the packets

$$\bar{n} = 115/81 = 1.42 \qquad (1.231)$$

Equation (1.229) gives the cost F of transmission on the network as a function of the mean transfer time. The corresponding curve is given in Figure 1.23. It can be seen that the network cost varies little when $\bar{T}$ is greater than 100 ms and increases very rapidly with $1/\bar{T}$ when $\bar{T}$ falls below 100 ms. The value $\bar{T} = 100$ ms will thus be chosen. Under these conditions, the various values of $C_u|_{\text{optimum}}$ are calculated using (1.228) which gives the results presented in Table 1.5. The transfer times $\bar{T}_u|_{\text{optimum}}$ on the various lines can then be evaluated using (1.216)

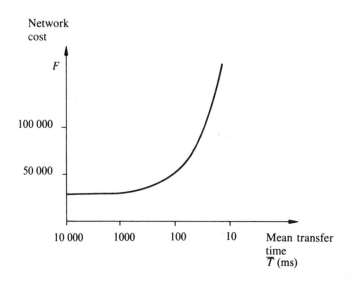

Figure 1.23 Network cost of Figure 1.22 as a function of mean transfer time

and are also given in Table 1.5. The mean transfer time on each network route is evaluated by adding the mean transfer times $\bar{T}_u$ on each of the lines which form the route. The traffic between Geneva and Zurich uses lines 1 and 2 on which the mean transfer times are 59 ms and 66 ms respectively. The mean transfer time between Geneva and Zurich is thus $59 + 66 = 125$ ms. The lines used by the various routes and the delay on each route are indicated in Table 1.6.

The method just described raises several points. Firstly, it can be seen that the capacities obtained can take any value. It was seen in Chapter 2 (Vol. I) that the capacities of digital lines are standardized to discrete values, for example 2400 bit/ s, 4800 bit/s, 7200 bit/s and 9600 bit/s for medium speed lines. The results obtained by the optimization method must therefore be rounded to the nearest standard value. Because of this, optimization can only be approximate.

Another important point concerns the criterion used for optimization of capacity allocation. It has been assumed so far that the designer would set the mean value $\bar{T}$ of all the transfer times for the set of routes and that he would attempt to minimize the cost of the network. Optimization can be performed, by a very similar method, by fixing the cost F of the network and attempting to minimize the transfer time $\bar{T}$. In practice, optimization of capacity allocation is often performed with criteria slightly different from cost or mean transfer time according to the techniques which will be described in the following two sections.

1.4.3 The various criteria for linear allocation of capacity

It has been seen in the previous section how the capacity allocated to various lines can be calculated in order to minimize the network cost for a fixed mean transfer time. This optimization method has the disadvantage of introducing large

Table 1.6 Mean transfer time on the various routes of the network of Figure 1.22 (mean of the set of mean transfer times: 100ms).

		Destination			
		1 Geneva	2 Lausanne	3 Zurich	4 Bern	5 Basle
Source	1 Geneva		1 58 ms	1–2 125 ms	1–4 129 ms	6 142 ms
	2 Lausanne			2 66 ms	4 71 ms	4–5 140 ms
	3 Zurich				3 79 ms	3–5 148 ms
	4 Bern					5 69 ms
	5 Basle					

differences between the mean transfer times on different routes, particularly among those where the traffic is small. If the example of the network of Figure 1.22 is considered again, the mean transfer time on the 'fastest' route (Lausanne–Geneva) is 58 ms while the mean transfer time on the 'slowest' route (Zurich–Basle) is 148 ms (Table 1.6), although the mean of the mean transfer times is 100 ms. It can very well happen that users are very satisfied with a transfer time of 58 ms and consider a transfer time of 100 ms to be still acceptable but cannot in any case accept a transfer time of 148 ms. To cater for this type of situation, it is necessary to be able to design a network in which the deviations $S(T)$ between the mean transfer times on the various paths are minimized; this leads to making the allocation for a given deviation $S(T)$ given by

$$S(T) = \left[\sum_{u=1}^{M} \frac{\lambda_u}{\bar{n}\gamma} (\bar{n}\bar{T}_u - \bar{T})^2 \right]^{1/2} \tag{1.232}$$

More generally [1.26], the optimization can be performed by looking for the capacity allocation which gives a minimum cost for a given value of $T_{(K)}$ where

$$T_{(K)} = \left[\frac{1}{\gamma} \sum_{u=1}^{M} \lambda_u (T_u)^K \right]^{1/K} = \left[\frac{1}{\gamma} \sum_{u=1}^{M} \lambda_u \left(\frac{1}{\mu_u C_u - \lambda_u} \right)^K \right]^{1/K} \tag{1.233}$$

For $K = 1$, the optimization is the same as that which was performed in the previous chapter and, by varying K, the optimization criterion is modified. For $K \neq 1$, the cost function F defined by (1.222) can be minimized by the method of Lagrangian multipliers:

$$F = \sum_{u=1}^{M} (a_u C_u + b_u) - \alpha \left[T_{(K)} - \left(\frac{1}{\gamma} \sum_{u=1}^{M} \lambda_u \left(\frac{1}{\mu_u C_u - \lambda_u} \right)^K \right)^{1/K} \right] \} \tag{1.234}$$

By using the same method as in the above section:

$$\frac{\partial F}{\partial C_u} = a_u - \frac{\alpha K \lambda_u \mu_u}{K\gamma(\mu_u C_u - \lambda_u)^{K+1}} \left[\frac{1}{\gamma} \sum_{u=1}^{M} \lambda_u \left(\frac{1}{\mu_u C_u - \lambda_u} \right)^K \right]^{\frac{1}{K}-1} = 0 \tag{1.235}$$

or

$$\frac{\partial F}{\partial C_u} = a_u - \frac{\alpha \lambda_u \mu_u T_{(K)}^{1-K}}{\gamma(\mu_u C_u - \lambda_u)^{K+1}} = 0 \tag{1.236}$$

and

$$\left(\frac{1}{\mu_u C_u - \lambda_u} \right)^K = \left(\frac{\gamma a_u}{\alpha \lambda_u \mu_u T_{(K)}^{1-K}} \right)^{K/(K+1)} \tag{1.237}$$

By multiplying the two terms of (1.237) by λ_u/γ and summing over all lines, the following is obtained:

$$T_{(K)}^K = \sum_{u=1}^{M} \frac{\lambda_u}{\gamma} \left(\frac{\gamma a_u}{\alpha \lambda_u \mu_u T_{(K)}^{1-K}} \right)^{K/(K+1)} \tag{1.238}$$

from which

$$\alpha^{K/(K+1)} = \sum_{u=1}^{M} \frac{\lambda_u}{\gamma} \left(\frac{\gamma a_u}{\lambda_u \mu_u T_{(K)}^2} \right)^{K/(K+1)} \tag{1.239}$$

and, using (1.236)

$$C_u \Big|_{\text{optimum}} = \frac{\lambda_u}{\mu_u} + \frac{1}{\mu_u} \left(\frac{\lambda_u \mu_u T_{(K)}^{1-K}}{a_u \gamma} \right)^{1/(K+1)} \left[\sum_{u=1}^{M} \frac{\lambda_u}{\gamma} \left(\frac{\gamma a_u}{\lambda_u \mu_u T_{(K)}^2} \right)^{K/(K+1)} \right]^{1/K} \tag{1.240}$$

or

$$C_u \Big|_{\text{optimum}} = \frac{\lambda_u}{\mu_u} + \frac{1}{\gamma^{1/K} T_{(K)}} \left(\frac{\lambda_u}{a_u \mu_u^K} \right)^{1/(K+1)} \left[\sum_{u=1}^{M} \left(\frac{a_u \lambda_u^{1/K}}{\mu_u} \right)^{K/(K+1)} \right]^{1/K} \tag{1.241}$$

It can easily be verified that (1.241) gives the same result as (1.228) for $K = 1$ and, for any value of K, this equation always consists of a term λ_u/μ_u to which a correcting term is added. λ_u/μ_u is the capacity which must be allocated to the channel for it to handle the traffic with an infinite mean transfer time. The correcting term corresponds to the additional capacity which must be given to the channel so that it can handle the traffic with the required transfer time.

By substituting the values of C_u defined by (1.241) into (1.222), the minimum cost F_{min} is obtained for the optimization criterion which corresponds to the chosen value of K:

$$F_{\text{min}} = \sum_{u=1}^{M} \left(\frac{a_u \lambda_u}{\mu_u} + b_u \right)$$

$$+ \frac{1}{\gamma^{1/K} T_{(K)}} \sum_{u=1}^{M} \left(\frac{\lambda_u a_u^K}{\mu_u^K} \right)^{1/(K+1)} \left[\sum_{u=1}^{M} \left(\frac{a_u \lambda_u^{1/K}}{\mu_u} \right)^{K/(K+1)} \right]^{1/K} \tag{1.242}$$

$$F_{\text{min}} = \sum_{u=1}^{M} \left(\frac{a_u \lambda_u}{\mu_u} + b_u \right) + \frac{1}{\gamma^{1/K} T_{(K)}} \left[\sum_{u=1}^{M} \left(\frac{a_u \lambda_u^{1/K}}{\mu_u} \right)^{K/(K+1)} \right]^{1/K} \tag{1.243}$$

When $K \to \infty$, Equations (1.233), (1.241) and (1.243) respectively become

$$T_{(K)} \Big|_{K\to\infty} = \sum_{u=1}^{M} \frac{1}{\mu_u C_u - \lambda_u} \tag{1.244}$$

$$C_u \Big|_{\substack{\text{optimum} \\ K\to\infty}} = \frac{\lambda_u}{\mu_u} + \frac{1}{\mu_u T_{(K)}} \tag{1.245}$$

$$F \Big|_{\substack{\text{min} \\ K\to\infty}} = \sum_{u=1}^{M} \left(\frac{a_u \lambda_u}{\mu_u} + b_u \right) + \frac{1}{T_{(K)}} \sum_{u=1}^{M} \frac{a_u}{\mu_u} \tag{1.246}$$

Equation (1.245) shows that optimization with $K = \infty$ reduces, when the message length $1/\mu_u$ is constant throughout the network, to assigning a capacity to each line which is the sum of a constant $1/\mu T(\infty)$ and a term λ_u/μ_u proportional to the traffic on the line. The mean transfer time $\bar{T}_u$ on each line and the mean

$\bar{T}$ of the transfer times on the whole network are obtained very simply for $K = \infty$ by substituting C_u defined by (1.245) into (1.216) and (1.221) which gives

$$\bar{T}_u \bigg|_{K=\infty} = T_{(K)} \tag{1.247}$$

$$\bar{T} \bigg|_{K=\infty} = \frac{\lambda}{\gamma} T_{(K)} \tag{1.248}$$

The mean transfer time is thus the same on each line when $K = \infty$.

In order to show the role of the parameter K in the optimization of capacity allocation, the example of the network of Figure 1.22 will be used again, with the same assumptions as in the previous section in respect of the traffic and message lengths, but with K varying between zero and infinity. The mean transfer times calculated under these conditions on the various routes are indicated in Table 1.7. It can be seen in the first four columns of the table that for a constant network cost of around 50 000, the mean transfer time $\bar{T}$ calculated for all routes of the network is a minimum ($\bar{T} = 100$ ms) for $K = 1$. The table also shows that the spread between the transfer times of the various routes of the network becomes smaller as K becomes larger. Furthermore, reference to the fifth column of the table shows that if a small increase in the cost of the network is permitted, it is possible to greatly reduce the spread of transfer times while keeping the mean transfer time of the whole network approximately constant. Hence, with the example used, the solutions for $K = 1$ and $K = \infty$ lead to costs of 50 094 and 55 040 respectively for mean transfer times of 100 ms and 85 ms. Under these

Table 1.7 Comparison of transfer times on the network routes of Figure 1.22 for various values of the optimization parameter K.

		K				
		0.5	1	2	∞	∞
Geneva–Lausanne	1	54 ms	58 ms	62 ms	74 ms	60 ms
Geneva–Zurich	1–2	118 ms	125 ms	130 ms	148 ms	120 ms
Geneva–Bern	1–4	125 ms	129 ms	133 ms	148 ms	120 ms
Geneva–Basle	1–6	178 ms	142 ms	113 ms	148 ms	120 ms
Lausanne–Zurich	2	64 ms	66 ms	68 ms	74 ms	60 ms
Lausanne–Bern	4	71 ms	71 ms	71 ms	74 ms	60 ms
Lausanne–Basle	4–5	139 ms	140 ms	141 ms	148 ms	120 ms
Zurich–Bern	3	82 ms	79 ms	77 ms	74 ms	60 ms
Zurich–Basle	3–5	150 ms	148 ms	147 ms	148 ms	120 ms
Bern–Basle	5	68 ms	70 ms	70 ms	74 ms	60 ms
Mean transfer time for the whole network $\bar{T}$		101 ms	100 ms	100 ms	105 ms	85 ms
Network cost F		50 082	50 094	50 166	50 196	55 040
$T_{(K)}$		140 ms	100 ms	85 ms	74 ms	60 ms

conditions the transfer time on the various routes varies only between 60 ms and 120 ms for $K = \infty$ compared with 58 ms and 148 ms for $K = 1$. These considerations thus lead to performing the optimization by taking a value of K close to 1 when minimization of cost is of prime importance and a value of K close to infinity when it is required to provide a service as similar as possible on the various routes.

1.4.4 Non-linear allocation of capacity

In principle, the allocation methods allow only a first approximation for the capacities to be provided on the links; this is because the cost of the latter is not, in general, a linear function of capacity but more of a stepwise function of the type shown in Figure 1.24. Furthermore, when lines from different telecommunication networks are used, the cost of a link no longer depends solely on its capacity and range but also on the particular network through which the link passes.

The simplest method to take account of the real cost of lines is to evaluate the linear approximation which is closest to the stepwise curve (Figure 1.24) and to apply the previous methods to calculate the capacities to be given to the lines. These capacities are then rounded to the standard discrete values (2400, 4800, 9600 bit/s, etc.) while checking that the latter still permit adequate traffic flow when they are adjusted to a value less than the calculated one.

Optimization can be performed more accurately by defining a continuous function whose general form is very close to the stepwise function. In this case, the optimization method remains very similar [1.28] to that which is used with a linear cost function. This technique is in fact rarely used since defining a suitable

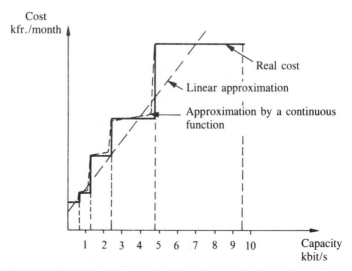

Figure 1.24 Link cost as a function of capacity for a given range

optimization function is difficult. In most cases therefore, non-linear optimization is performed from stepwise functions using heuristic procedures [1.29], [1.30] whose principle is as follows.

In the first step, capacity allocation is made from linear approximations to the stepwise functions with rounding of the calculated values to the nearest discrete values. To refine the capacity allocation, a number of programs are used which determine if it is profitable to either increase or decrease the capacity of the various lines by one increment. These programs determine the effect of the capacity of each link on the transfer time of the paths used by the network and the relative cost of an increase or decrease of capacity. It is thus possible to optimize the network by successive attempts at increasing the capacity of lines for which a minimum increase in cost leads to a maximum reduction in transfer time, and decreasing the capacity of lines which have the opposite effect. This optimization technique does not guarantee an optimum result, but in general it gives good results and can be applied to complex problems, for example the allocation of capacity in a network containing packets which belong to different priority classes.

1.4.5 Optimization of the topology of transit networks

For a transit network having a fixed topology, routing techniques permit the paths followed by the various packets to be determined and capacity allocation methods allow the capacities of the network links to be chosen as a function of cost and delay criteria. In most cases, the topology is not fixed and the problem to be solved is to determine the best possible topology for a given traffic configuration. In practice, it is generally accepted that the geographical location of the nodes is known, which is reasonable taking account of the fact that the nodes form the entry points to the transit network. Even with this simplification, the problem of determining the links which must be provided to carry the traffic under the best possible conditions is very complex; in practice it is only partly solved by means of heuristic methods [1.31], [1.32]. All these methods start from a given topology and modify it by a series of perturbations; each time, routing and capacity allocation methods are used to compute the cost of the modified network while adopting those modifications which reduce the cost or delay.

In its simplest form, the perturbation method proceeds by *branch exchange* and consists of selecting two adjacent links S_1S_2 and S_3S_4 which operate between four distinct nodes which are S_1, S_2, S_3 and S_4 in this case (Figure 1.25). The two links S_1S_2 and S_3S_4 are then dropped and replaced by two other links, S_1S_3 and S_2S_4 in this case, which interconnect the same nodes. The cost of the network, modified in this way, is evaluated by applying the routing and capacity allocation methods, and this new topology replaces the previous one if its cost is less. The exchange algorithm applies particularly well to links of high cost and low capacity, as in the case where the introduction of a direct link between nodes is justified by the magnitude of the direct traffic between these two nodes.

Topology optimization can be carried out in a more global and more efficient manner by using a method [1.32] of decoupling the network into two sub-networks

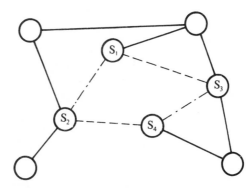

Figure 1.25 Optimization of the topology by exchanging lines

(*saturated cutsets*) intended to reduce the traffic on the most heavily loaded lines of the network. The method involves the following steps:

Step 1   Decoupling into two sub-networks

The various links are ordered according to their rate of use, that is their traffic intensity (traffic/capacity of the link). The links are cut one by one, starting with the link whose traffic intensity is the highest and continuing in decreasing order of traffic intensity until the network becomes split into two independent sub-networks (Figure 1.26). All the links which have been cut are then re-examined and those which do not reconnect the two sub-networks are re-established. At the end of this operation, the decoupling is minimal since it concerns all the links between the two sub-networks and only these links.

The algorithm continues by adding or dropping links in accordance with the following rules.

Step 2   Adding a link

This procedure aims to link the two sub-networks again while reducing the traffic on the links which were previously cut and were overloaded. With this aim, the algorithm provides for addition of the least costly link which connects the two sub-networks. In order to unjam the previously cut links, the two nodes which define the new link must be separated by at least two links from the nodes which define the cut between the two sub-networks. Thus, with the example of Figure 1.26, the network can be modified by inserting the S_2S_4 link.

Step 3   Dropping a link

The dropping procedure is intended to remove from the network the link which is the most costly and whose rate of use is the lowest. The various links are

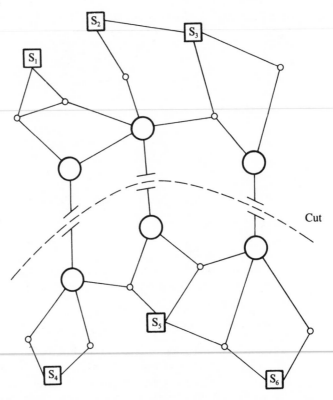

Figure 1.26 Optimization of the topology of a transit network by the cutting method

evaluated according to a criterion G_i which takes account of these two factors by using

$$G_i = f_i \frac{(C_i - \lambda_i/\mu_i)}{C_i} \qquad (1.249)$$

where f_i and C_i are the cost and capacity of the link respectively and λ_i/μ_i is the mean traffic, expressed in bit/s, which flows in the link. The algorithm eliminates the link in the network for which G_i is the highest.

Optimization of the network is achieved using the preceding procedures while checking, each time a link is removed or added, whether the resulting network has a better or worse performance (with respect to cost and delay) than the previous one. The method is terminated when it has not been possible to obtain any further significant improvement after a certain number of iterations. Elimination of uneconomic links must be performed with care, since it tends to reduce the connectivity of the network, that is its reliability. This sometimes leads to extending the method by including a program to calculate the connectivity and check that this remains within tolerable limits after elimination of the link under consideration.

1.4.6 Capacity allocation in a centralized local network

The user's equipment, which is often a terminal, is connected to nodes of the transit network by local networks which generally have a tree structure (Figure 1.27). The problems posed by the design of these networks are in principle the same as those encountered in connection with transit networks, but the solutions are generally simpler because of the centralized structure which is used. Local networks connect a number of pieces of user's equipment S_i, which will be called the set of terminals for the sake of simplicity, to a node G_0 of the transit network. The terminals S_i can be connected to G_0 either directly or by way of concentrators G_j. With such a network, routing of packets does not pose a problem, since there is one and only one path between G_0 and each of the terminals. In contrast, it is necessary, as before, to determine the capacity to be allocated to each of the links and to define the topology of the network. The latter problem, which will be tackled in the following sections, reduces to determining where the concentrators should be located, defining which terminals must be connected to which concentrators and specifying how the terminals are connected to the concentrators.

Capacity allocation will first be considered by showing that the non-linear allocation problem can be solved in a relatively simple manner by making use of the tree structure of the network. From now on, we will make use of a simplified representation for the network, as shown in Figure 1.28. As the topology of the network is known, the routing is also known and it is easy to compute the traffic λ_i on the various links from the traffic γ_j between the terminals and the central node G_0, which is part of the problem data. For the example of the network of Figure 1.28, the traffic entering is given by Table 1.8 and the traffic on the various lines is indicated in Figure 1.28. The cost f_i of the lines is generally quoted

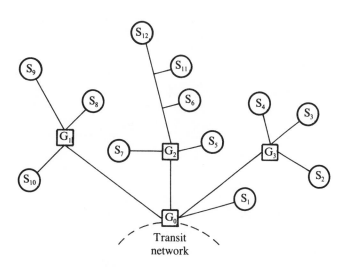

Figure 1.27 Local tree network

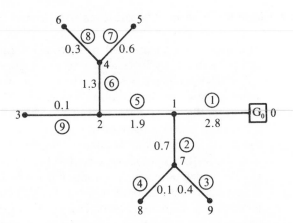

Figure 1.28 Simplified representation of a tree network

as a monthly rental, which is a function of line length and capacity C, in a table of the type given in Table 1.9. Knowing the range of the various lines, it is then easy to compute the cost of each line as a function of capacity which gives a table of the type represented in Table 1.10 for the network used as an example.

If the traffic is Poisson with an exponential message length distribution, the mean transfer time $\bar{T}_i$ on the various network lines can be calculated using the now familiar equation

$$\bar{T}_i = \frac{1}{\mu_i C_i - \lambda_i} \tag{1.250}$$

The example with packets of constant length $1/\mu$ of 256 bits results in Table 1.10 which provides the mean transfer times as a function of the capacity of the lines.

The capacity allocation will now be sought which minimizes the maximum transfer time on the network while retaining the lowest possible cost. For this

Table 1.8 Traffic entering the network of Figure 1.28.

Node j	1	2	3	4	5	6	7	8	9
Input traffic γ_j	0.2	0.5	0.1	0.4	0.6	0.3	0.2	0.1	0.4

Table 1.9 Rental cost of lines as a function of capacity and range (arbitrary values).

Capacity C (bit/s)	600	1200	2400	4800	9600
Cost f fr./month/km	10	12	18	20	30

Table 1.10 Costs and delays on the various lines as a function of capacity (network of Figure 1.28), $1/\mu = 256$.

Link	Traffic λ_i/μ_i (bit/s)	Transfer time (ms) cost (fr./ month)	Line capacity (bit/s)				
			600	1200	2400	4800	9600
① 10 km	716.8	$\bar{T}_i$ f_i	∞ 100	530 120	152 120	63 200	29 300
② 5 km	179.2	$\bar{T}_i$ f_i	608 50	251 60	115 90	55 100	27 150
③ 20 km	102.4	$\bar{T}_i$ f_i	514 200	233 240	111 360	55 400	27 600
④ 12 km	25.6	$\bar{T}_i$ f_i	446 120	218 144	108 216	54 240	27 360
⑤ 30 km	486.4	$\bar{T}_i$ f_i	2253 300	359 360	134 540	59 600	28 900
⑥ 2 km	332.8	$\bar{T}_i$ f_i	958 20	295 24	124 36	57 40	28 60
⑦ 8 km	153.6	$\bar{T}_i$ f_i	573 80	245 96	114 144	55 160	27 240
⑧ 15 km	76.8	$\bar{T}_i$ f_i	489 150	228 180	110 270	54 300	27 450
⑨ 7 km	25.6	$\bar{T}_i$ f_i	446 70	218 84	108 126	54 140	27 210

purpose, an algorithm will be used which reduces the network to a simpler equivalent network by means of a series of parallel and serial merges.

As the transfer time can be a maximum only for packets originating from a station located at one of the ends of the network, the parallel and serial merge algorithms are applied from the ends of the network with progression towards the centre G_0. At the first step, the links from the ends which converge to the same node are considered and a table is established for each group of such links; this provides the maximum transfer time as a function of the cost of the link. Hence, for the network of Figure 1.28, the links from ends ⑦ and ⑧ meet at the same node. From Table 1.10, if a capacity of 600 bit/s is chosen for link ⑦, its cost is 80 fr. and the delay is 573 ms. As the delays corresponding to all possible capacities on line ⑧ are less than 573 ms, the lowest possible capacity, that is 600 bit/s, can be assigned to line ⑧ in this case without risk of increasing the maximum transfer time. Hence line ⑧ in this case has a cost of 150 fr. which gives a total cost for the two lines of $80 + 150 = 230$ fr. for a delay of 573 ms. This gives the first element of the cost–delay table for the two merged links ⑦ and ⑧ (Table 1.11). Passing to the immediately greater capacity on line ⑦ (1200 bit/s in this case), the delay (245 ms in this case) is compared with the corresponding delays on the other line, starting with the highest delays. The

Table 1.11 Optimum delay–cost relation after parallel merge of links ⑦/⑧ and ③/④.

Merged links	Maximum transfer time $\bar{T}_i$ (ms) cost f_i (fr./month)									
⑦ + ⑧	$\bar{T}_i$	573	489	245	228	114	110	55	54	27
	f_i	230	246	276	324	414	430	460	540	690
③ + ④	$\bar{T}_i$	514	446	233	218	111	108	55	54	27
	f_i	320	360	384	504	576	616	640	840	960

highest delay on line ⑧ corresponds to 489 ms for a capacity of 600 bit s. As the delay on line ⑧ is higher than that which corresponds to 1200 bit/s on line ⑦, the maximum transfer time is 489 ms for this combination and the corresponding cost is 246 fr. In this way there is a subsequent element in the table which depicts the parallel merge ⑦/⑧ and, by continuing, the values given in Table 1.11 are obtained. The same procedure is used for all the parallel links from the ends (③ and ④ in this case), which leads to the reduced network of Figure 1.29.

To continue the reduction of the network, it is now necessary to perform serial merges, for example merge line ⑥ with lines ⑦ + ⑧. In the case of a serial merge, the delays and costs must be added and the procedure is as above, retaining only the transfer times which correspond to a minimum cost. After serial merging of (⑦ + ⑧) with ⑥ and (③ + ④) with ②, the network is reduced to four branches as shown in Figure 1.30 and the characteristics corresponding to the merged branches reduce to those of Table 1.12. Simplification of the

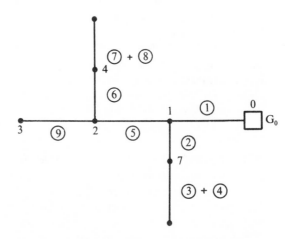

Figure 1.29 Reduced network after a first series of parallel merges

Table 1.12 Delay/cost table after the first merges.

($\overline{7}$ + $\overline{8}$) + $\overline{6}$	$\bar{T}_i$	868	630	546	302	256	171	142	83	55
	f_i	254	270	286	316	384	454	474	520	750
($\overline{3}$ + $\overline{4}$) + $\overline{2}$	$\bar{T}_i$	765	569	484	288	260	245	138	82	54
	f_i	380	420	444	484	534	654	726	790	1110

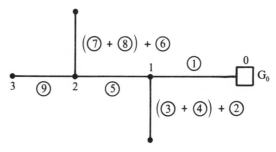

Figure 1.30 Reduced network after the first merges

network continues with a succession of parallel and serial merges until it is reduced to a single branch. This produces (Table 1.13 and Figure 1.31) a series of transfer times, with the corresponding costs, which permit the capacity allocation which minimizes the maximum transfer time to be chosen for a given cost. If, for example, a maximum cost of 1448 is chosen, the maximum transfer time is 678 ms and, if care has been taken to note the capacity allocations which correspond to various delays at each stage of merging, it is easy to determine the capacity allocation of the network for the maximum value of delay which has been chosen. The capacity allocation of the network is given by Figure 1.32 for the case of a maximum transfer time of 678 ms. Notice, in this example, that the transfer times corresponding to nodes at the ends are relatively close, which is logical since the capacity allocation has been made so as to minimize the cost for a given maximum transfer time.

Table 1.13 Maximum transfer time as a function of the cost of the network for an optimum capacity allocation.

$\bar{T}_i$	1757	1519	1435	1335	1141	1052	968	868
f_i	1184	1200	1216	1246	1260	1280	1296	1326
$\bar{T}_i$	828	724	678	640	632	547	499	423
f_i	1340	1380	1448	1518	1560	1584	1624	1684
$\bar{T}_i$	378	344	317	306	289	264	230	205
f_i	1692	1792	1922	1972	2014	2126	2172	2186
$\bar{T}_i$	196	171	167	143	140	140	112	
f_i	2272	2286	2516	2580	2580	2650	2880	

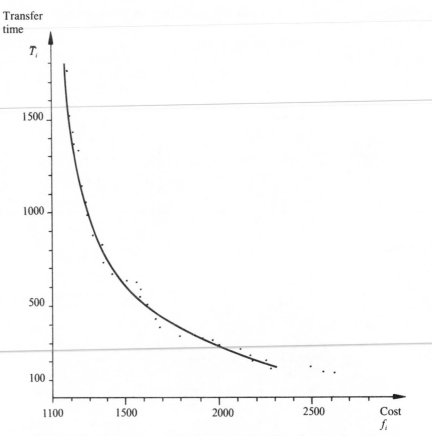

Figure 1.31 Maximum transfer time as a function of network cost

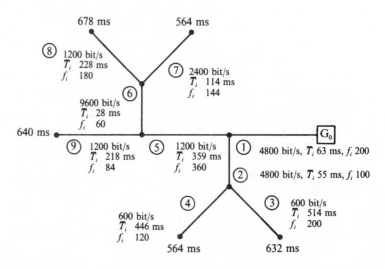

Figure 1.32 Optimum capacity allocation for a maximum transfer time of 678 ms

1.4.7 Selection of concentrators

The problem which arises here is to determine where to locate the concentrators and which terminals must be connected to which concentrators (Figure 1.27). It will be assumed that the local network contains N terminals S_i which can be connected to the central node G_0 either directly or through one of the M concentrators G_j of the network. If f_{ij} is the cost of a connection between the terminal S_i and the concentrator G_j, the total cost of the network is given by

$$F = \sum_{i=1}^{N} \sum_{j=0}^{M} x_{ij} f_{ij} + \sum_{j=0}^{M} y_j g_j \qquad (1.251)$$

where g_j is the cost of the concentrator G_j and its connection to G_0; x_{ij} and y_j are binary variables which indicate the presence or otherwise of a link or a concentrator respectively, where

$$x_{ij} = 1 \text{ if } S_i \text{ and } G_j \text{ are connected}$$

$$= 0 \text{ if } S_i \text{ and } G_j \text{ are not connected} \qquad (1.252)$$

$$y_j = 1 \text{ if the concentrator } G_j \text{ exists}$$

$$= 0 \text{ if the concentrator } G_j \text{ does not exist} \qquad (1.253)$$

The goal here is to minimize the total cost F of the network; it is understood that a terminal S_i can be connected to only one concentrator which implies that

$$\sum_{j=0}^{M} x_{ij} = 1, \quad i = 1, \ldots, N \qquad (1.254)$$

It is generally assumed that, with the exception of the central node G_0, all the concentrators have a capacity limited to L lines to link with terminals, which introduces the additional constraint

$$\sum_{i=1}^{N} x_{ij} \leq L, \quad j = 1, \ldots, M \qquad (1.255)$$

The problem of minimization of total cost F defined by (1.251), taking account of constraints (1.254) and (1.255), is generally solved by assuming that the geographic locations of the M concentrators and N terminals are known (Figure 1.33) together with the costs g_i and f_{ij} (Table 1.14). The topology can be simply defined from this data by using heuristic algorithms of which the best known are the add and drop algorithms described below.

1.4.8 Add algorithm

The first stage of the add algorithm consists of directly connecting all the terminals to the central node G_0 (Figure 1.34). The cost F^1 of this topology, which does not include any concentrator other than the central node, is evaluated very simply from the cost matrix f_{ij} of the links (Table 1.14) with

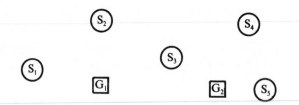

Figure 1.33 Diagram of terminals and concentrators

Table 1.14 Cost of concentrators and links.

Concentrators G_j					G_1	G_2
Cost of concentrators G_j (including the cost of the concentrator–central node link)					2	2

Terminals	S_1	S_2	S_3	S_4	S_5	
Concentrators						Matrix of
G_0	1	3	2	4	2	link costs
G_1	1	2	1	3	3	f_{ij}
G_2	4	3	1	2	1	

$$F^1 = \sum_{i=1}^{N} f_{i0} \tag{1.256}$$

In the case of the network whose costs are defined by Table 1.14, it follows that $F^1 = 12$. An attempt will now be made to reduce the cost by introducing one, and only one, concentrator. To do this, successive attempts are made to

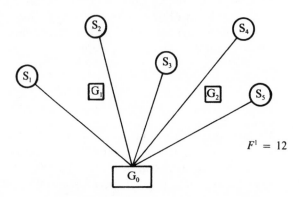

Figure 1.34 Add algorithm: first stage

introduce each of the possible concentrators by assigning to this concentrator the stations for which connection to the concentrator represents the greatest possible savings. The concentrator is finally retained, which permits the greatest cost reduction of the network with respect to F^1, clearly on condition that such a solution exists. A new and more efficient topology is thus obtained whose cost is F^2 and which includes a concentrator with

$$F^2 = g_j + \underset{\substack{i \\ \text{direct} \\ \text{links}}}{\sum} f_{i0} + \underset{\substack{i \\ \text{links} \\ \text{passing} \\ \text{through } G_j}}{\sum} f_{ij} \qquad (1.257)$$

With the proposed example and assuming that the number of input ports L to the concentrators is unlimited, for the concentrator G_i,

$$F_1^2 = 2 + (1 + 2) + (2 + 1 + 3) = 11 \qquad (1.258)$$

Similarly, if a concentrator G_2 is added instead of concentrator G_1, the cost of the network F_2^2 becomes

$$F_2^2 = 2 + (1 + 3) + (1 + 2 + 1) = 10 \qquad (1.259)$$

As $F_2^2 < F_1^2 < F^1$, the concentrator G_2 is definitely added which leads to the network of Figure 1.35 whose cost F_2^2 is 10.

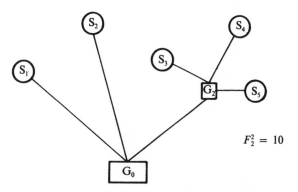

$$F_2^2 = 10$$

Figure 1.35 Add algorithm: second stage

The procedure is repeated by attempting at each step to introduce an additional concentrator into the network in the same manner. At each new step, an attempt is made to connect all the terminals to the new concentrator, without considering previous connections. Hence, in the present example, an attempt can be made to introduce the concentrator G_1. This gives a new cost $F_{1,2}^3$ with

$$F_{1,2}^3 = 4 + 1 + (2 + 1) + (2 + 1) = 11 \qquad (1.260)$$

In this case $F_{1,2}^3 > F_2^2$ which shows that adding an additional concentrator increases the cost instead of reducing it. The algorithm is therefore terminated while retaining the network obtained at the end of the preceding stage.

1.4.9 Drop algorithm

The procedure adopted by the drop algorithm is exactly the reverse of that which corresponds to the add algorithm. The first step consists of locating the M concentrators in the network and determining to which concentrators the various terminals should be connected to achieve the minimum network cost F^1. This stage is very simply evaluated from the cost matrix f_{ij} (Table 1.15) by choosing the line connected to the concentrator for which the cost is minimum for each column, that is for each terminal. For the example, this leads to the network of Figure 1.36 whose cost is given by

$$F^1 = (2 + 2) + 1 + 2 + 1 + 2 + 1 = 11 \qquad (1.261)$$

Attempts are then made to withdraw one, then two, then M concentrators from the network, looking each time for the concentrator whose elimination permits the maximum cost reduction; the algorithm terminates when dropping a concentrator increases the cost instead of reducing it.

Proceeding in this way with the example of Figure 1.36, it is easily shown that the concentrator G_1 must be eliminated and concentrator G_2 should be retained. This finally leads to the topology of Figure 1.35 which is the same as that obtained with the add algorithm. When the networks are complex, the results obtained are not necessarily the same with the two algorithms, since the latter do not

Table 1.15 Allocation of terminals to concentrators in the first stage of the drop algorithm.

Terminals	S_1	S_2	S_3	S_4	S_5
Concentrators					
G_0	①	3	2	4	2
G_1	1	②	①	3	3
G_2	4	3	1	②	①

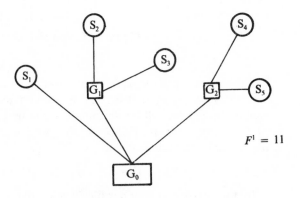

Figure 1.36 Drop algorithm: first stage

guarantee a minimum cost solution. In practice, the two algorithms give good results in most cases.

1.4.10 Optimum topology with multipoint lines

So far, it has been assumed that the local network consists only of concentrators and point-to-point lines. In many cases, the local traffic is relatively small and it is then advantageous to connect the terminals to the network by way of multipoint lines. The problem of global optimization of a local network containing multipoint lines has not been satisfactorily solved at the present time, but it is possible to obtain an optimum solution when the location of the concentrators is known and the terminals are assigned to the different concentrators. It will therefore be assumed that the choice of concentrators and assignment of terminals are made using the methods described in the previous sections, and attention here will be devoted to the problem of the optimum connection of a specified number of terminals S_i to a central site S_0 by means of point-to-point or multipoint lines (Figure 1.37). This problem corresponds equally well to the connection of terminals to concentrators as to the connection of concentrators to the input node of the transit network. For simplicity, it will be assumed that all lines of the network have the same capacity C which will be assumed to be 2400 bit/s in the example considered here.

The traffic produced by the terminals S_i towards the central node S_0 is λ_i/μ_i (Table 1.16) and the connection costs between the various stations are defined by a cost matrix f_{ij} which is given for the case of the present example in Table 1.17.

It is now required to determine the minimal cost tree which enables the terminals to be connected to the central site S_0. If there were no constraint due to the maximum capacity C on the lines, the minimum tree could be determined very easily. In this case it is only necessary to start by connecting the station whose connection cost is minimum, S_4 in this case, to S_0. The remaining station which is closest to either S_0 or S_4 is then found (Table 1.17) and connected to

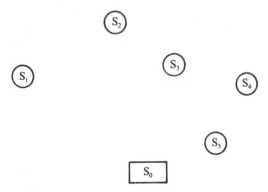

Figure 1.37 Diagram of the terminals to be connected to the central node S_0 by multipoint lines

Table 1.16 Traffic produced by the stations (bit/second).

Station S_i	S_0	S_1	S_2	S_3	S_4	S_5
Traffic produced by stations λ_i/μ_i (bit/s)	/	600	900	1000	200	1600

Table 1.17 Matrix of connection costs between the terminals.

Terminal	S_0	S_1	S_2	S_3	S_4	S_5
S_0	/	5	7	8	2	3
S_1	5	/	5	3	6	7
S_2	7	5	/	5	10	8
S_3	8	3	5	/	5	3
S_4	2	6	10	5	/	3
S_5	3	7	8	3	3	/

the network. In this case, the remaining station closest to S_0 is S_5 with a connection cost of 3 and the remaining station closest to S_4 is also S_5 with a connection cost also of 3. Hence S_5 can be connected to either S_4 or S_0 but it is clearly preferable to connect S_5 to S_0 to avoid overloading line S_0S_4. Proceeding step by step in this way finally gives the unconstrained minimum cost tree which is shown in Figure 1.38 for the case of the present example. The solution obtained in this way is, in general, not acceptable since it leads, most of the time, to traffic which exceeds the maximum authorized capacity on some lines. It can be seen, for example, that the traffic on lines S_3S_5 and S_5S_0 is 2500 bit/s and 4100 bit/s respectively; this is greater than the capacity of the lines, which is 2400 bit/s. The

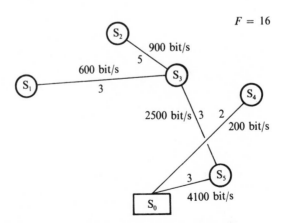

Figure 1.38 Optimum multipoint network without capacity constraints

network obtained in this way is useful, however, since it provides a lower limit F_{min} of the cost of a realizable network and it serves as a starting point for optimization of the real network. Notice that F_{min} is given here by

$$F_{min} = 3 + 5 + 3 + 2 + 3 = 16 \qquad (1.262)$$

The solution with capacity constraints is obtained by starting from the observation that the links directly connected to the central node S_0 in the unconstrained solution remain connected to S_0 in the constrained solution [1.33]. This gives a network with only the direct connections to S_0 of the unconstrained network (Figure 1.39) as a starting point for the solution.

The algorithm continues by dividing the set of possible links into two subsets of which one includes any additional link and the other forbids the possibility of using this same link. In this way only one link is introduced at this stage which does not lead to the maximum capacity constraint being exceeded. With this example, the additional link $S_3 S_4$ can be introduced into the first subset (Figure 1.40) and forbidden from inclusion in the second subset (Figure 1.41); all other

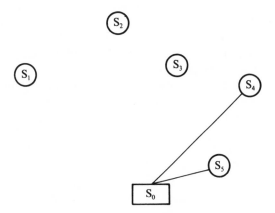

Figure 1.39 First stage of the optimization of a multipoint network with capacity constraints

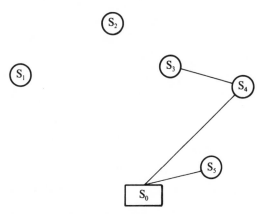

Figure 1.40 Second stage of the optimization of a multipoint network with capacity constraints: first subset

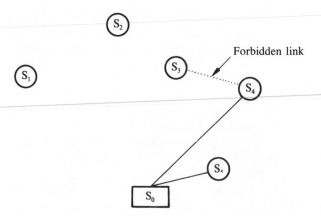

Figure 1.41 Second stage of the optimization of a multipoint network with capacity constraints: second subset

links are permitted. The unconstrained optimization algorithm is then applied to the two subsets and the costs of the two corresponding networks are compared. The network whose cost F^1 is the smallest is selected and checked against the capacity constraints. If the constraints are satisfied, the network thus obtained is the optimum network with constraints. Otherwise, the cost F^1 is a new lower cost limit for the network and the subset of minimum cost is itself divided into two subsets. The algorithm continues until a realizable network is obtained.

In the present case, the first two subsets, corresponding to Figures 1.40 and 1.41, both lead to costs of 18 and an overload on some links. One or the other of the two subsets may be chosen to be again divided into two subsets. The subset of Figure 1.40 will be taken here and this is divided into the two subsets shown in Figures 1.42 and 1.43. The unconstrained optimization algorithm is applied to the two new subsets which again gives two networks of the same cost

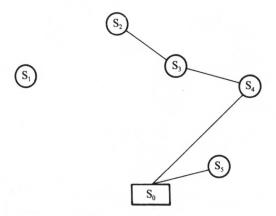

Figure 1.42 Third stage of the optimization of a multipoint network with capacity constraints: first subset

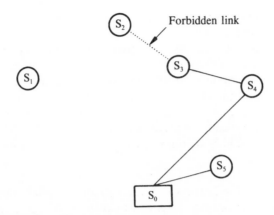

Figure 1.43 Third stage of the optimization of a multipoint network with capacity constraints: second subset

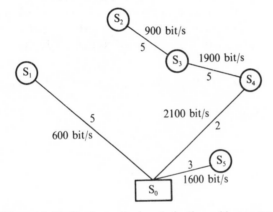

Figure 1.44 Final result of optimization with constraints

which violate the capacity constraint. The algorithm continues in an iterative manner until the subset leading to a minimum cost also gives a realizable network. In the case of this example, the network obtained in this way has a cost of 20 and its topology is given in Figure 1.44.

It has been assumed here that the nodes of multipoint lines are located at the station locations. This condition is not mandatory and the algorithm applies in a similar manner when the nodes of multipoint lines are separate from the stations. In this case, the locations of the stations are in general defined by their cartesian coordinates and the variables are the distances between the stations and the tree rather than the distances between stations.

1.4.11 Heuristic design algorithms for multipoint networks

The method presented in the previous section guarantees that the multipoint network obtained is of minimum cost. This algorithm has the disadvantage of requiring substantial computation. Because of this, it is often useful to use much simpler heuristic algorithms, which generally allow networks to be obtained whose cost is only slightly greater than that of an optimum network. One of these algorithms will be described here, the Esau–Williams algorithm [1.35]. This algorithm compares the cost of connecting each station to the central node S_0 and to all other nodes. The stations for which the difference in connection cost is the greatest are connected first while checking at each stage that the capacity constraints are satisfied. The cost differences are represented by the cost index a_{ij} which is computed for all values of i and j from

$$a_{ij} = f_{ij} - f_{i0} \qquad (1.263)$$

A positive value for a_{ij} indicates that the connection cost of node S_i to the central node S_0 is less than the cost of connecting S_i to node S_j.

At the first step of the algorithm, all the a_{ij} are computed from the cost matrix f_{ij}. This gives a table such as Table 1.18 which is computed here from the cost matrix of Table 1.17. At the second step the node S_u is chosen whose index a_{uv} has the highest negative value and it is connected to the corresponding node S_v. In the present case, the highest negative value is $a_{3,1} = -5$ and S_3 is therefore connected to S_1. A check is then made that the capacity constraints are satisfied.

Table 1.18 Esau–Williams algorithm. Matrix of indices of relative costs.

Terminal	S_1	S_2	S_3	S_4	S_5
S_1	/	0	−2	1	2
S_2	−2	/	−2	3	1
S_3	−5	−3	/	−3	−5
S_4	4	8	3	/	1
S_5	4	5	0	0	/

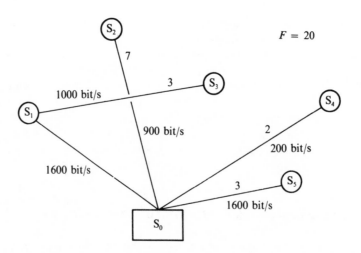

Figure 1.45 Network produced by the Esau–Williams algorithm

If they are, which is the case here since the sum of the traffic of S_3 and S_1 (1000 bit/s + 600 bit/s) remains less than 2400 bit/s, S_u is definitely connected to S_v and the node whose index a_{ij} now has the highest negative value is examined. When the capacity constraints are not satisfied, a_{ij} is given the value $+\infty$, to indicate that the corresponding link is impossible, and the following node is examined. With this procedure, the network of Figure 1.45 is obtained for the example where the total cost is 20. The network obtained is optimum in this case. For more complicated networks, a sub-optimum solution is generally obtained whose cost is usually greater than that of the optimum network by 5 to 10 per cent. Similar results are obtained with other heuristic algorithms such as Prim's algorithm [1.34].

CHAPTER 2

NETWORK
LAYER

2.1 INTRODUCTION

Direct communication between stations is provided by data links which ensure error-free transmission of frames (Vol. I, Chap. 4). When the number of stations to be connected becomes large, the traffic can no longer be expedited on a single channel and it is then necessary to provide the communication system with a network structure which contains a number of nodes A, B, C, D, . . . , interconnected by data links (Figure 2.1). Once the data exchanged by the stations has to pass through one or more nodes, the design and operation of the communication system raises a series of new problems compared with those arising in a simple data link.

Matters relating to definition of the network topology, resource allocation and multiplexing have already been examined in the previous chapter. It will therefore

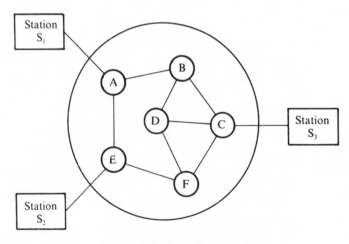

Figure 2.1 Network example

be assumed that the network topology is known and routing of packets on the network will be considered here.

From the user's point of view, the network must permit an exchange of messages between subscribers. At this level, which corresponds to the *transport layer* of the OSI model, the size of messages must be determined only by the needs of the users and it must not, therefore, depend on constraints associated with transport of the messages. In general terms, the transport layer manages the end-to-end transport of the messages and provides independence from higher-level layers with respect to all problems of this type. Transport is realized in the form of **packets** which are routed between two stations connected to a communication network. This allows an intermediate layer to be defined between the transport and data link layers. This intermediate layer is called the **network layer** and its main role is to establish a path between the two nodes that connect the network users. The most important functions provided by the network layer are the segmentation of *messages into packets* whose size is adapted to the characteristics of the network, together with **addressing** and **routing** of the packets and network **flow control**. The network layer thus makes the transport layer independent of the problems of routing and switching.

In order to clarify these concepts, the network of Figure 2.1 will be considered again. It will be assumed that user S_1 transmits messages to user S_3. Communication must, therefore, be provided between nodes A and C. If the messages sent by S_1 are too long, they must be segmented into several packets of a size better suited to effective utilization of the network. There are several possible routes through the network between A and C. The network layer services must, therefore, choose a free path to route the packets from A to C, without necessarily selecting the same path for successive packets. The packets are routed from node to node in the form of frames defined by the line procedure. Each network node is equipped with output buffers to store the packets waiting for transmission on the multiplexed output lines. The packets arrive at the destination node C with a variable delay and can even be delivered in a sequence different from that of transmission, since they can have followed different paths and some of them can have been retransmitted following errors on the lines. It is, therefore, necessary for the packets to be put back into sequence by node C or by station S_3 before being reassembled and delivered in the form of messages to user S_3. At a given time, the traffic offered by the users can exceed the network throughput; if the situation is not controlled, this produces an overload of the node buffers and a catastrophic degradation of the network performance. The network services must, therefore, ensure flow control and load balancing to prevent this **congestion**.

It can be seen that messages are routed in the form of packets on a network, consisting of nodes equipped with data processing hardware capable of temporarily storing packets and transmitting them in the form of frames on channels which they manage using a line procedure. In practice, the network can be public or private. In the case of a private network, the nodes are installed and managed by the user; they are interconnected with channels leased from the public network and whose sole function is the transmission of bits, that is the provision of the services of the physical layer of the OSI model. In contrast, a public network provides a set of network functions, and the interface with the user is generally located at the origin and destination nodes, that is between the network layer and the transport layer. An example of this latter approach with X.25 protocol will be given in this chapter.

Because of the large variety of network types, the distribution of functions between the network and transport layers can vary from one network to another. As far as the functions of the network layer are concerned, attention here will be devoted to segmentation and reassembly of messages together with the problems of routing and congestion. This chapter concludes with a presentation of the CCITT X.25 network protocol and the ISO network protocols [2.1]–[2.44].

2.2 SEGMENTATION AND REASSEMBLY OF MESSAGES

Messages sent by users can vary widely in size according to the type of application and the nature of the message. A reply to a request can be very short and even reduce to one bit in the case of an unqualified acknowledgement. In contrast, some messages can be very long, for example when they correspond to a file transfer. On the other hand, effective utilization of a network requires the transmitted packets to be neither too long nor too short. Each packet contains the text and a header which includes, for example, the destination address, an indication of the packet type and an error check word. If the packets are too short, the relative size of the text with respect to the header is small and the lines and node buffers are poorly used. If, on the other hand, the packets are too long, the header becomes negligible with respect to the text, which is favourable from the standpoint of line utilization. However, for a given bit error rate, the probability that a packet is in error and must be retransmitted becomes greater as the packet becomes longer. As seen in Chapter 3 (Vol. I) there is an optimum packet size for transmission on a line with a given error rate.

Segmentation and reassembly of messages permits the line utilization rate to be optimized and consequently the cost of communication to be reduced. This procedure also reduces the system response time, that is the message transfer time in the network. While passing from node to node, the packets suffer a delay which is proportional to their length, since error checking requires that they can be retransmitted only after having been completely received; this means that they must be temporarily stored at each node. More precisely, if the messages are not segmented into packets, the transfer time T_R of N network nodes (the input node plus $N - 1$ intermediate nodes) is given by

$$T_R = N(T_t + T_w) \qquad (2.1)$$

where T_t is the transmission time of a message on a channel (all channels are assumed to be identical) and T_w is the waiting time at a node, that is the delay between arrival at a node of the last bit of a message and the start of transmission to the following node (Figure 2.2). If each message is now segmented into M packets, transmission can be performed *with overlap* by retransmitting the first packets of the message on the following channel before the arrival of the last packets at the node (Figure 2.2). The delay T_w corresponds essentially to the waiting time in the output queue of the node. If T_w is the same when the messages

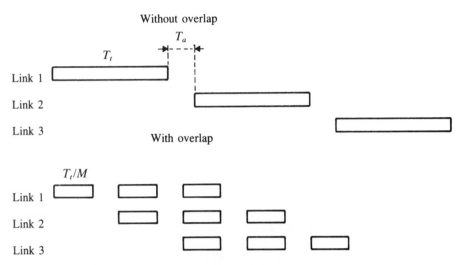

Figure 2.2 Transmission on successive links of a network

are transmitted as such, or segmented into packets, the transfer time T_R, in the case of transmission with overlap, is given by

$$T_R = T_t + (N - 1)\left(\frac{T_t}{M} + T_w\right) + M\,T_w \tag{2.2}$$

This clearly shows that the transfer time can be appreciably reduced if T_w is small with respect to T_t and if the size of the headers is negligible with respect to the text.

The effect of packet length on the system response time can be analyzed more precisely by calculating the transfer time of the output queue of each node. In the simple case where each node is modelled by an M/M/1 queue, with an arrival of λ packets per second at the node and an output capacity of μ packets per second, the mean transfer time of the node by the packet is $(1/\mu)[1/(1 - \rho)]$, with $\rho = \lambda/\mu$. This delay is equal to $2/\mu$ for a traffic intensity of ρ of 0.5 and reaches $5/\mu$ for $\rho = 0.8$. It can be seen that the delay at each node is of the order of two to five times the transmission time $T_t = 1/\mu$ on the output channel. This clearly indicates that this delay becomes smaller as T_t becomes smaller, that is the packets are shorter.

In principle, the network and transport layers will group short messages and segment long messages so that transmission is performed with packets whose length is close to an optimum which is a function of the required cost and response time.

2.3 DATAGRAMS AND VIRTUAL CIRCUITS

The services provided by the network at the transport layer can be either of the datagram type or the virtual circuit type.

With the *datagram* service, users at the transport level hold a dialogue by means of an exchange of packets which are treated as independent entities by the network. The transport layer must, therefore, provide the destination address, and generally its own address, each time it sends a packet. Datagrams are routed individually within the network, in the same way as telegrams in a conventional communication network. It can be seen that, in this case, the network provides a minimum service, with a 'best effort' routing of the packets which are entrusted to it. As successive datagrams can follow different paths, the network cannot guarantee to deliver them in sequence to the destination transport layer. Furthermore, the network does not provide any *flow control* to match the capacity of the transmitter to that of the receiver.

The network can also provide a complete service in the form of *virtual circuits* whose mode of operation is similar to that of the telephone switched network. With such an approach, the dialogue between two correspondents divides into an establishment phase, a text transfer phase and a release phase. During the establishment phase, the calling subscriber sends a request for connection with the called party. This request causes establishment of a path, called a virtual circuit, on which all the packets exchanged between the correspondents circulate until communication is broken off. It can be seen that, as for conventional telephone communication, the calling subscriber provides the address of the called user only at the start of the dialogue. The text is then transmitted in the form of packets carrying the number of the virtual circuit used and no longer the source and destination addresses as in the case of a datagram. Each node contains a table of active virtual circuits which indicates to which output channel the incoming packets must be routed. These tables are updated during the opening and closing of virtual circuits and map a virtual circuit number to the address of the adjacent node. To prevent conflicts when several users choose the same virtual circuit number, each node performs a conversion of the virtual circuit number by making an input virtual circuit number, associated with the address of the node located upstream, correspond with an output virtual circuit number associated with the address of the node located downstream. In this way each node can assign a free circuit number to the virtual circuit under consideration during the establishment phase.

The network operates by packet switching both when the service is of the datagram type and when the service is of the virtual circuit type. However, the network layer provides a more complete service to the transport layer in the case of a virtual circuit service. The network provides the transport layer with packets in sequence and it performs error monitoring by ensuring that no packets are lost. The role of the transport layer can then be reduced to providing recovery in the case of network failure. The datagram service, in contrast, requires the transport layer to take charge of end-to-end error monitoring together with resequencing of messages.

The relative usefulness of a datagram service compared with a virtual circuit service depends very much on the type of application. As datagrams must contain the complete address of the destination, the size of the packet headers is in general larger than with a virtual circuit; this reduces transmission efficiency when the stations exchange short messages. In this case, it is preferable to use virtual circuits. Furthermore, when the virtual circuit service is provided by a public network, the users' task is greatly simplified, since most of the communication functions are provided by the public network. Datagrams are preferable to virtual circuits, mainly for some real-time applications which require a short response time and when users can tolerate some lost or out-of-sequence packets. In this case, a virtual circuit procedure would only complicate network supervision and delay the delivery of messages without real benefit to the subscribers. Similarly, when the network is itself connected to other networks, it is preferable to manage the interfaces between the networks in the form of datagrams to avoid needless cascading of several error and sequence checking services.

Long-haul networks in principle always provide a virtual circuit service. In contrast, some local area networks such as MAP (*Manufacturing Automation Protocol*) are designed for datagram operation which is better suited to this type of environment where the packets are routed on short channels with almost perfect sequencing.

In practice, the network layer can provide, at the interface with the transport layer, a service of a different type from that which is used internally. An often used combination is to operate the network by datagrams and to present the user with an interface of the virtual circuit type. Notice finally that virtual circuit networks generally offer the additional facility of establishing *permanent virtual circuits* whose use is comparable to that of a special line in a conventional telephone network.

2.4 ROUTING ALGORITHMS

2.4.1 Overview

One of the main problems which must be solved at the network level is to establish a path between the calling and the called stations. When the network is operated by datagrams, path selection must be made separately for each packet; in contrast, the decision on the route to be followed is taken only at the time of establishment for virtual circuits. In the two cases, the choice of routing algorithm is not easy since it must satisfy a large number of often contradictory requirements. This algorithm should be simple, in order to be implemented easily in the nodes, and it should ensure correct routing of packets for any hazards suffered by the network. The algorithm must thus be capable of giving satisfactory results in spite of variations in traffic and the topology of the network. It must also ensure approximately equal sharing of access rights to the network among the various stations. Finally, the routing algorithm must, if possible, permit operation of the network in an optimum manner, according to a criterion which can vary with the type of use. In most cases, the goal is realization of a network which minimizes the packet transfer time or which maximizes network throughput. The principal objectives are thus minimum transfer time and maximum throughput. In other cases, the main objective of the designer is, for example, to reduce the cost of communication or develop a reliable network which is capable of surviving catastrophic failures and whose performance does not excessively degrade when the traffic peaks.

In view of the large number of possible objectives, there is a large number of different types of routing algorithms [2.1]–[2.3]. These can correspond to *deterministic or adaptive policies* according to whether or not they adapt to variations of traffic and network topology. Also, routing algorithms can be *centralised* if the paths are defined by one particular node. In the other case, the routing algorithm is *distributed* between all the nodes; this is favourable from the point of view of reliability but complicates the algorithm and makes optimization of packet routing more difficult. Notice finally that there are algorithms which are more difficult to categorize and which use a **random routing** or **routing by flooding** technique.

2.4.2 Routing by flooding

The technique of *routing by flooding* [2.3]–[2.4] is based on a simple approach which consists of retransmitting the packets at each node on all the output links of the node with the exception of the incoming link. A node connected to *K* other nodes thus retransmits $K - 1$ copies of the packet which it has just received, as indicated in Figure 2.3. It can be seen that the flooding technique ensures the delivery of at least one copy of the packet at the destination provided there is at least one path between it and the source of the packet. This guarantee of routing is ensured even if the topology changes, for example after a catastrophic failure of certain network components. The flooding technique is therefore robust, which explains why its main applications are in military networks. Also, as all possible links between the source and destination nodes are tried exhaustively, the method ensures that at least one of the copies of the packet will reach the destination by

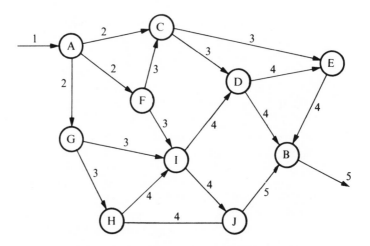

Figure 2.3 Operation of a network by flooding

the shortest path and hence with a *minimum delay* if the network is lightly loaded. The flooding technique also has the advantage of being very simple to implement, since message routing requires neither routing tables nor knowledge of the geographical location of the destination on the network. For the packet to reach its destination, it is merely necessary that it is capable of recognizing that it is addressed to it.

The price to be paid for these qualities of simplicity, robustness and speed of routing is clearly poor utilization of the network resources and a marked tendency to congestion. Since each node in general produces several copies of the same packet, the total number of copies in transit on the network increases very rapidly as a function of the number of nodes passed through. It is therefore necessary to limit the number of packets in transit and to eliminate the copies still being routed after the first version of the packet has reached its destination. To eliminate the redundant packets, the destination can return, by flooding, an acknowledgement message after the arrival of the first copy of a packet. This acknowledgement reaches various nodes of the network which can then proceed to destroy the copies which have become useless. This method has the disadvantage of producing substantial additional traffic to purge the packets. It is in general more advantageous to use a special field in the packet header which indicates the maximum number of nodes which a packet is allowed to pass through. This field is decremented each time the packet passes through a node and every packet containing a zero field is purged. This guarantees that the packets have a limited lifetime and cannot in any case circulate indefinitely within the network. Another possibility for reducing the network load consists of numbering the packets sequentially. In this case, the nodes retransmit only the packets which they have never received and which they recognize as such by their source address and sequence number. Packets which return to a node which they have already passed through are thus destroyed. This method ensures quick elimination of expired packets at the cost of more complex processing at the nodes, since these must maintain a table of the packets which pass through.

As a consequence of the considerable dummy traffic which is produced by the network, the flooding technique is reserved for extreme situations where the objectives of reliability and simplicity outweigh all other considerations. It is, however, possible to retain some advantages of the flooding technique while making better use of network resources; the method used – random routing – is derived from it. On the other hand, recent studies have shown that routing by flooding is well suited to the implementation of meshed local area networks of high capacity [2.5].

2.4.3 Random routing

The technique of *random (or stochastic) routing*, like the flooding method, has the characteristic of not requiring the nodes to know the structure of the network, or the state of the traffic, in order to take routing decisions. However, the nodes in this case avoid sending copies of the received packets systematically on all the output links in order not to produce excessive dummy traffic. The random routing technique consists of sending one or more replicas of the received packet on output links which are chosen either randomly or according to scant information on the general direction followed by the packet. In the latter case, the routing method is called *selective flooding*.

With the simplest random routing method, each node retransmits the packet on one of the output channels chosen at random. With a suitably meshed network, the packet always arrives at its destination even if the path which it follows is intricate. It can be seen that this policy leads to a very simple implementation of routing at node level and it avoids overload caused by packet duplication. In contrast to these advantages, the routing delay is on average longer than with deterministic techniques since the packets tend to follow intricate routes instead of taking the most direct route to their destination.

The performance of a random routing network can be evaluated in a much simplified manner by assuming a homogeneous network consisting of N nodes each connected to K adjacent nodes by a direct line [2.3]. It will be assumed that each node produces a mean of λ packets per unit time with a Poisson arrival of packets and an exponential packet length distribution with a mean length $1/\mu$. The unit of time considered here will be called a *cycle* and it will be assumed that each link between nodes has a capacity of μ packets per cycle. Under these conditions, the mean number of packets $M(t)$ present in the network at the end of a cycle is equal to the number of packets $M(t-1)$ present in the network at the end of the preceding cycle plus the number of packets produced during a cycle minus the number of packets which arrived at their destination. The set of packets produced during a cycle is clearly λN. If the network load is small, the packets pass through μ links on average each cycle. The number of packets at the destination during a cycle is thus equal to $M(t-1)\mu p$, where p is the probability of a packet arriving at its destination. Hence

$$M(t) = M(t-1) + \lambda N - \mu p\, M(t-1) \tag{2.3}$$

with

$$p = p_1 p_2 \tag{2.4}$$

where p_1 is the probability that the packet is located at a node adjacent to the destination node, and p_2 is the probability that the adjacent node sends the packet to its destination. As each node is connected to K other nodes, in the case of purely random routing, clearly

$$p_1 = K/N \tag{2.5}$$

$$p_2 = 1/K \tag{2.6}$$

$$p = 1/N \tag{2.7}$$

The probability p is thus independent of time t. By assuming that the initial condition is $M(0) = 0$, the solution to Equation (2.3) is given by

$$M(t) = \frac{\lambda N}{\mu p} [1 - (1 - \mu p)^t] \tag{2.8}$$

In the steady state, that is for $t \to \infty$, the number of packets $M(\infty)$ in the network reduces to

$$M(\infty) = \lambda N / \mu p \qquad (2.9)$$

or, using (2.7)

$$M(\infty) = \lambda N^2 / \mu \qquad (2.10)$$

In order to evaluate the mean network transfer time $\bar{T}_R$ expressed as a number of cycles, notice first that the time spent by a packet in going from one node to the following one is equal to the transmission time on the line, or $1/\mu$, plus the waiting time in the queue at the node of departure. Assuming an M/M/1 queue, the mean waiting time is $\rho/\mu(1 - \rho)$ with $\rho = \lambda/\mu$. The mean time of passing through a node to the following one $\bar{T}_p$ is thus given by

$$\bar{T}_p = \frac{1}{\mu}\left(1 + \frac{\rho}{1 - \rho}\right) = \frac{1}{\mu(1 - \rho)} \qquad (2.11)$$

It has been indicated above that p is the probability that a packet located at any node of the network is routed to its destination after a time interval $\bar{T}_p$. The probability that the packet remains waiting in the network for a time $i\bar{T}_p$ is thus equal to $(1 - p)^i p$, which leads to a mean number $\bar{i}$ of waiting intervals given by

$$\bar{i} = \sum_{i=1}^{\infty} ip\,(1 - p)^i = \frac{1 - p}{p} \qquad (2.12)$$

and, using (2.11)

$$\bar{T}_R = \bar{i}\,\bar{T}_p = \frac{1 - p}{\mu p(1 - \rho)} \qquad (2.13)$$

When the network is lightly loaded, $\rho \ll 1$. In this case, the mean transfer time for purely random routing reduces to

$$\bar{T}_R = \frac{N - 1}{\mu} \simeq \frac{N}{\mu} \qquad (2.14)$$

It is now easy to understand why random routing leads to long transfer times. Assume that the nodes have some information on the final destination of the packet and the nodes adjacent to the destination node know the line to which it is connected. In this case, if N is large and the nodes are distributed randomly in the network, the number of nodes connected to the destination by two links in cascade is $K(K - 1)$. The probability p_1 that a packet is located at a node adjacent to the destination node is thus given by

$$p_1 = \frac{K(K - 1)}{N} \times \frac{1}{K} = \frac{K - 1}{N} \qquad (2.15)$$

with $p = p_1$ in this case, since the node adjacent to the destination node sends it the packet directly.

Still assuming light traffic, this leads to a network transfer time which is reduced approximately by a factor $1/(K - 1)$:

$$\bar{T}_R \simeq \frac{N}{\mu(K-1)} - \frac{1}{\mu} \qquad (2.16)$$

It can be seen that even scant knowledge of the route to be followed considerably reduces the packet routing time and permits better utilization of the lines. These are the reasons why most networks make use of the routing techniques which will be examined below and which are based on knowledge of the network topology and the geographical location of the correspondents.

2.4.4 Fixed routing method

The **fixed routing** method [2.3], [2.6]–[2.8] defines the paths which the various packets must follow from the general characteristics of the network such as the topology and the mean traffic anticipated on the transmission channels. The routing rules are in principle established once and for all and aim to optimize the performance criteria selected by the designer. In most cases, network optimization aims to minimize the mean packet transfer time. The fixed routing technique requires a routing table at each node; this table permits the node to determine on which output channel it must send a packet which it has just received in order to route it to its destination under the best possible conditions. The fixed routing method is thus simple in principle, since the only processing performed at the nodes amounts to management of the queues for the channels and reading of the tables; consequently the route optimization algorithm is operated only when the network is designed. In practice, the routing tables cannot be completely fixed, since the topology of the network can change following equipment failure or because of the introduction of new users. It is, therefore, necessary to provide techniques for updating the tables to take account of these situations. In this case fixed routing will be understood to mean a routing method in which the routing tables are updated only rarely; this contrasts with *adaptive routing* in which the routing tables are continuously updated, for example as a function of the present state of the traffic. It can be seen that the fixed routing method aims for long-term global optimization, while the adaptive method has the goal of satisfying the optimality criterion at all times.

With fixed routing, the mean packet transfer time is significantly shorter than with random routing. To show this, the network model examined in the previous chapter [2.3] can be considered again. In this case, fixed routing implies that the fraction p of packets reaching their destination in the following cycle is equal to the inverse of the mean path length. For a homogeneous circular network the mean path length is approximately equal to the radius of the circle and the number of nodes is proportional to the square of the radius; p is thus given by

$$p = 2/\sqrt{N} \qquad (2.17)$$

Using (2.9) and (2.13), the number of packets $M(\infty)$ present in the network in the steady state and the mean network transfer time $\bar{T}_R$ become

$$M(\infty) = \lambda N^{3/2} / 2\mu \qquad (2.18)$$

$$\bar{T}_R \simeq \sqrt{N} / 2\mu \qquad\qquad (2.19)$$

This shows that the number of packets in transit in the network and the transfer time are reduced by a factor of approximately $2\sqrt{N}$ if random routing is replaced by an optimum fixed routing.

In order to specify the main features of fixed routing, the simple case of a network in which a path between two nodes is considered to be optimum if it passes through a minimum number of nodes will be considered first. If the example of Figure 2.3 is considered again, two optimum paths exist between A and B which are ACDB and ACEB. To organize routing in the network, it is first necessary to determine the $N(N-1)$ optimum paths between all pairs of nodes. In practice, the links are generally bidirectional, so the total number of optimum paths reduces to $N(N-1)/2$. Once all the optimum paths are established, the network designer determines a routing table containing $N-1$ lines for each node. This table indicates the adjacent node to which a packet addressed to a given node must be routed to follow the optimum path. By way of example, Table 2.1 represents the routing table of node C in the case of the network of Figure 2.3 with the criterion of the minimum number of nodes passed through. There are often several minimal paths between pairs of nodes. In this case, it is advantageous to balance the traffic between these different paths by establishing multiple choice routing tables (Table 2.2) which indicate the different adjacent nodes to which the packets can be routed for each destination. Each output is thus assigned a weighting factor, and the choice of the particular direction in which a packet is sent is made by a weighted random choice.

Determination of the optimum paths is greatly facilitated by using the very general optimality principle which specifies that every optimum path can consist only of optimum segments. This principle is very simply proved by noting that if an optimum path contains a non-optimum segment, the latter can be replaced by a shorter segment which reduces the length of the optimum path and is thus contrary to the assumption of optimality of the path. One of the consequences

Table 2.1 Routing table of node C.

Destination address of the incoming packet	Address of the adjacent node to which the packet must be routed
A	A
B	D
C	–
D	D
E	E
F	F
G	A
H	A
I	F
J	F

Table 2.2 Multiple choice routing table. Node C.

Destination address of the incoming packet	Addresses of the adjacent nodes and probabilities of choice					
A	A	1				
B	D	0.5	E	0.5		
C						
D	D	1				
E	E	1				
F	F	1				
G	A	1				
H	A	0.33	F	0.33	D	0.33
I	F	0.5	D	0.5		
J	F	0.33	D	0.33	E	0.33

of the optimality principle is that the set of optimal paths with a node as destination forms a tree whose root is the destination node (Figure 2.4). This tree is built very simply by placing all the nodes adjacent to the destination at the first level; all nodes which are not already placed and are adjacent to first-level nodes are then placed at the second level and so on until all are placed.

So far only the simple case of a network where all links are identical has been considered. In practice, links generally have different costs and capacities. Furthermore, the mean traffic is not the same in all links and it cannot in any case exceed the maximum authorized capacity on the link. The problem for the network designer is to determine the routing tables as a function of the topology of the network and the traffic produced by the stations. These routing tables

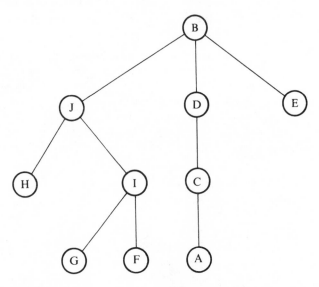

Figure 2.4 Tree of minimum paths to a destination node B

must be defined to minimize a 'cost' criterion globally for the whole network; the criterion can, for example, be the network transfer time or the cost of communication. In the following discussion the 'cost' and 'distance' of a link between two nodes i and j will be characterized by the variable $d(i, j)$ which is small when the 'cost' of the link is itself small; the minimum path between two nodes will be defined as the path along which the sum of the $d(i, j)$ is minimum. The problem in the design of routing tables is to determine the paths between all pairs of stations in such a way that the set of paths has a minimum or almost minimum cost. Before considering the question of global optimization of network routing, the problem of determining the minimum path between two stations will be considered first; this can be solved very simply using graph theory.

Consider any network, such as that of Figure 2.5(a) where each link is weighted by a cost value $d(i, j)$ indicated on the corresponding link of the graph. The minimum path between two stations such as A_0 and A_8 can be determined by exhaustively examining all possible paths between A_0 and A_8. The cost of each path is the sum of the $d(i, j)$ corresponding to the links forming this path and the costs of the various paths can be compared to determine the minimum path. This method leads very quickly to substantial computation when the number of nodes becomes large. To simplify computation, use is made of a *minimum path algorithm* [2.9] which exploits the optimality principle described previously.

The minimum path algorithm, starting with the source node (A_0 in this case), progressively establishes the minimum path between each node and the source node. This is achieved by indicating, for each node j, the address A_i of its neighbour located upstream on the minimum path together with the minimum distance $D(j)$ between the node and the source. When all the nodes are marked, the address A_i, contained in the label of the destination, points towards the node immediately upstream which itself points to the previous one and so on. The minimum path is thus established by successive pointers starting from the destination. The algorithm operates in the following manner:

Initialization: All nodes A_i are numbered in any order with the exception of the origin node and the destination node which are A_0 and A_{N-1} respectively for a graph containing N nodes. A label $(A_i, D(i))$ is assigned to each node and is initialized to $(\cdot, 0)$ for the origin node and $(\cdot, \infty)$ for all other nodes (initially the nodes are assumed to be not connected). The 'dot' symbol indicates that the address of the predecessor is not known. The distance $D(0)$ of the origin node to itself is zero. The distance between the origin node and all other nodes is infinite.

Marking of nodes starting from the origin: Let $i = 0$ and let j vary between 1 and $N - 1$. For every node A_j such that $D(i) + d(i, j) < D(j)$, replace the previous label of A_j by the label $(A_i, D(i) + d(i, j))$. Repeat the previous process by successively putting $i = 1, \ldots, N - 2$ and varying j between 0 and $N - 1$ each time. If in the course of these operations a link exists for which the index j of the final end is less than the index i of the initial end, put $i = j$ and restart the operations starting with j.

Determination of the minimum path: Start with the destination node A_{N-1}. If the label A_{N-1} does not contain the address of a predecessor, a path between the

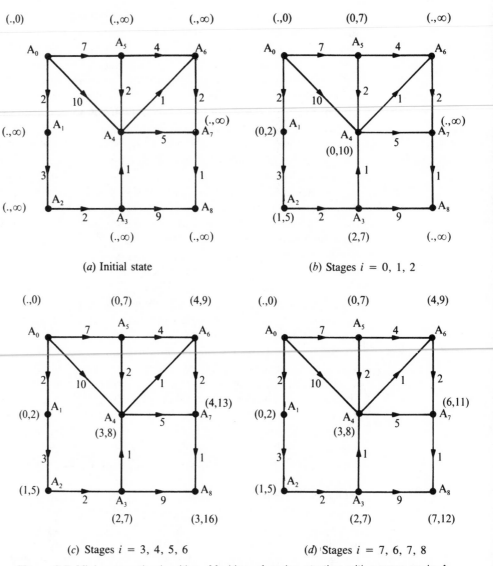

(a) Initial state

(b) Stages $i = 0, 1, 2$

(c) Stages $i = 3, 4, 5, 6$

(d) Stages $i = 7, 6, 7, 8$

Figure 2.5 Minimum path algorithm. Marking of nodes starting with source node A_0

origin and the destination does not exist. If the label contains the address of a node A_K, this is the predecessor of A_{N-1} on the minimum path. Continue the algorithm until the origin node is reached by taking the address of the predecessor in the label each time. The algorithm terminates when it arrives at the origin node.

The processes of marking and determination of the minimum path are illustrated in Figures 2.5 and 2.6 respectively. The minimum path algorithm can be mechanized very simply, as indicated by the flowchart of Figure 2.7. It is

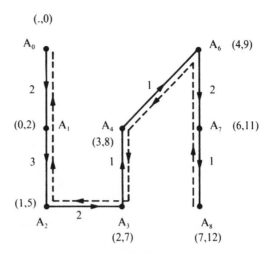

Figure 2.6 Minimum path algorithm. Determination of minimum path

convenient to represent the links with their weights in a table which shows the mapping between predecessor and successor nodes and contains the cost $d(i, j)$ for the cases corresponding to links which exist in the network (Figure 2.8). It can be shown that the algorithm described above computes the minimum path with a number of operations which is proportional to $N\log_2 N$.

It would seem that computation of routing tables reduces essentially to determining the minimum path for each source/destination pair K separately. In practice, this solution is not satisfactory since the cost functions $d(i, j)$ associated with the links depend in general on the traffic between the M pairs of the network $(K = 1, \ldots, M)$. If $d(i, j)$ represents the mean transit time on the link $A_i A_j$, the minimum path for a single pair will pass through the high-capacity channels of the network. However, if a large number of pairs share these high capacity channels they tend to overload which increases the corresponding $d(i, j)$. To solve the problem of fixed routing, it is therefore necessary to use a global optimization method which uses the minimum path algorithm in an iterative manner. The *flow deviation (FD) algorithm* [2.7] will be described here; this permits global minimization of the network packet transfer time.

It will be assumed that the network consists of N nodes and B links and the traffic entering the network will be defined by the mean number γ_{ij} of packets per second produced at node A_i and sent to node A_j. The total traffic γ entering the network is thus

$$\gamma = \sum_{ij} \gamma_{ij} \tag{2.20}$$

It will also be assumed that the arrival of packets on link B_i of capacity C_i bit/ s follows a Poisson law with a mean of λ_i packets per second and the packet length is distributed exponentially with a mean of $1/\mu$ bits per packet. With these assumptions, the mean packet transfer time $\bar{T}_R$ of the network is obtained very simply by calculating the sum of the transfer times t_i of each link, weighted as a

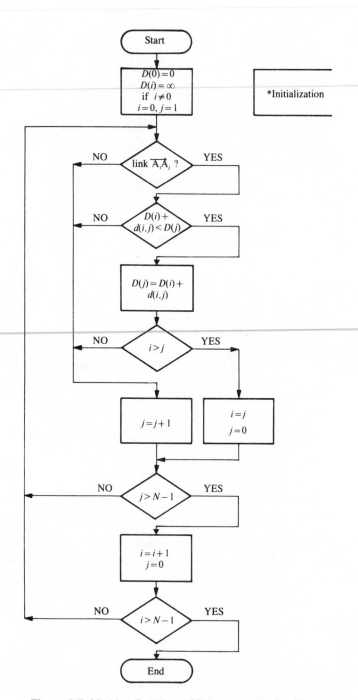

Figure 2.7 Marking flowchart. Minimum path algorithm

Node A_j (successor)

	A_0	A_1	A_2	A_3	A_4	A_5	A_6	A_7	A_8
A_0		2			10	7			
A_1			3						
A_2				2					
A_3					1				9
A_4							1	5	
A_5					2		4		
A_6								2	
A_7									1
A_8									

Node A_i (predecessor)

Figure 2.8 Table of links between the network nodes of Figure 2.5

function of the number of packets arriving on the link and referred to the total number γ of packets arriving in the network. Hence

$$\bar{T}_R = \frac{1}{\gamma} \sum_{i=1}^{B} \lambda_i t_i \tag{2.21}$$

The transfer time t_i of a link consists of a fixed propagation time T_i' and a delay $(\lambda_i/\mu)/(C_i - \lambda_i/\mu)$ which corresponds to the transmission time on the line and the waiting time in the queue. Hence, using $T_i = \mu T_i'$,

$$\bar{T}_R = \frac{1}{\gamma} \sum_{i=1}^{B} \left(\frac{f_i}{C_i - f_i} + f_i T_i \right) \tag{2.22}$$

where $f_i = \lambda_i/\mu$ is the mean data rate (flow) on link B_i which is clearly equal to the sum of the data rates f_i^K corresponding to each of the M source–destination pairs:

$$f_i = \sum_{K=1}^{M} f_i^K \tag{2.23}$$

On the other hand, the law of flow conservation requires that the traffic entering node i and corresponding to a source–destination pair must be equal to the traffic leaving the node and corresponding to the same pair. Finally the flow on a line must not exceed the link capacity C_i.

The FD algorithm determines, by successive iterations, the vector $\boldsymbol{f} = [f_1 f_2 \ldots f_i \ldots f_B]$ of the flows in the various links which minimizes $\bar{T}_R$ taking account the constraints indicated above. The algorithm is initialized to a value $\boldsymbol{f}(0)$. At iteration n, the flow values corresponding to the vector $\boldsymbol{f}(n)$ serve to calculate the relative incidence $d_i(n)$ of a flow variation in link i on the mean transfer time $\bar{T}_R$ where

$$d_i(n) = \frac{\partial \bar{T}_R}{\partial f_i(n)} = \frac{1}{\gamma}\left(\frac{C_i}{[C_i - f_i(n)]^2} + T_i\right) \tag{2.24}$$

At each iteration, the minimum path algorithm is used to determine the vector $g(n)$ of flows which minimizes the transfer time of the network, by taking the value $d_i(n)$ calculated using (2.24) as the 'distance' on each link. The new estimate $f(n + 1)$ of f is thus given by

$$f(n + 1) = f(n) + \alpha[g(n) - f(n)] \tag{2.25}$$

with

$$0 < \alpha < 1 \tag{2.26}$$

The algorithm is terminated when

$$f(n + 1) - f(n) < \epsilon \tag{2.27}$$

where ϵ is a vector defining the required approximation in the estimates of flow on the various lines.

The FD algorithm thus permits the mean transfer time of the network to be minimized for a specified topology and mean traffic. When these parameters vary very slowly, it is possible to have some adaptivity by assigning the task of periodically updating the routing tables as a function of the variations in traffic and topology to one of the stations of the network. This method, which is simple in principle, is called *centralized routing*. Its principal disadvantages are to make proper operation of the network depend on that of a single station and to cause substantial service traffic. The various stations must report the state of the network as they perceive it (adjacent nodes operational, traffic, etc.) to the routing centre and the centre must in turn send their new routing tables to the stations. The service traffic is thus concentrated around the routing station which tends to overload this part of the network. Finally, centralized routing can pose difficult compatibility problems at the time of updating the tables, since these cannot arrive at different stations at the same time. Because of this, some stations operate temporarily with old routing tables while others are already using the new ones. It is necessary to prevent this situation leading to the loss of packets or overloading of the network. This problem can be solved by increasing the updating delay but, in this case, the adaptation speed becomes very low.

To overcome these disadvantages, establishment of routing tables can be performed at each node in terms of the information available locally. This leads to distributed adaptive methods [2.10] which can be classified into local routing techniques and distributed routing techniques.

2.4.5 Local routing

Local and distributed routing methods are both based on the establishment for each node of its own routing table based on the information which it can obtain locally. For example, if the optimization criterion of the network consists of minimizing the network packet transfer time, each node A_i establishes an estimate $\tilde{T}_i(A_j, B_l)$ of the transfer time of path A_iA_j along the output line B_l of node A_i as a function of the information available locally. By repeating the estimate for all nodes A_j of the network, node A_i establishes a *table of estimated delays* (Table 2.3) which gives the estimated tranfer times as a function of the destination node and the output line. The *routing table* is thus derived very simply from the table of estimated delays by choosing the output line giving the shortest delay for each destination (Table 2.3).

With the local routing method, each node establishes its table of estimated delays solely from information which is directly available without exchanging information with its neighbours. In its simplest form, the local routing algorithm involves each node in trying to dispose of the packet which it has just received as quickly as possible by placing it in the shortest output queue (*the hot potato algorithm*) [2.11]. The node thus estimates the transfer time to the other nodes by the length of the queues on the output lines. The result is that the table of delays reduces to one line of which the various elements represent the length of the queues on the output lines.

The simple local algorithm tends, because of its principle, to cause packets to circulate quickly in the network. However, as the routes are chosen in an arbitrary manner, the mean path length is far from minimal so that the mean transfer time $\bar{T}_R$ is high. It is possible to improve this situation by combining local routing with static routing, for example by operating the node with static routing as long as the queues on the output channels do not exceed a certain length and changing to local routing when they do. Another possibility consists of weighting the choice of output channel from the node as a function of an index which combines the length of the path to the destination and the size of the output queue.

The performance of the local routing algorithm can be significantly improved by adopting a *backward learning* technique [2.11]–[2.14] which is based on the assumption that the transfer time from node A_i to node A_j is approximately the same as the transfer time from node A_j to node A_i (symmetrical traffic). Under these conditions, if each packet contains the address of the source node and the

Table 2.3 Methods of local and distributed routing. Delay table and routing table.

Destination nodes	Output lines of node A_i			Destination nodes	Output line of node A_i
	B_1	B_4	B_7		
A_1	0.4	0.2	0.3	A_1	B_4
A_2	0.1	0.6	0.2	A_2	B_1
A_i					
A_N	0.6	0.4	0.7	A_N	B_4

Table of estimated delays $\tilde{T}_i(A_j, B_l)$ Routing table time at which it left it, each node A_i can compute an estimate $\tilde{T}_i(A_j,B_l)$ of the transfer time to node A_j from the difference ΔH between the time of arrival and the time of departure of the packets which reach it from A_j by line B_l. In practice, $\tilde{T}_i(A_j,B_l)$ is established in an iterative manner by calculating a new estimate $\tilde{T}_i^{n+1}(A_j,B_l)$ from the former estimate $\tilde{T}_i^n(A_j,B_l)$ and the transfer time ΔH of the packets recently arrived from A_j where

$$\tilde{T}_i^{n+1}(A_j,B_l) = \tilde{T}_i^n(A_j,B_l) + \alpha[\Delta H - \tilde{T}_i^n(A_j,B_l)] \qquad (2.28)$$

where α is a constant between 0 and 1.

It can be seen that this very simple method has its equivalent in everyday life when a person departing for a given town asks travellers from this town about their travelling time.

As with most local and distributed methods, the backward learning technique leads to a system which quickly learns good news but reacts slowly to bad news. If new channels open in the network or existing channels transmit packets more rapidly because they are suddenly less loaded, the ΔH values corresponding to the packets received by A_i and passing through these channels will be suddenly smaller; this will rapidly lead A_i to reduce the estimates of the corresponding $\tilde{T}_i(A_j,B_l)$. If, in contrast, some nodes or lines of the network are subject to a fault, A_i will cease to receive packets passing through these network components and this will lead to the current values of $\tilde{T}_i(A_j,B_l)$ being retained, in spite of the fact that the corresponding links are cut. In a strongly meshed network, the system will slowly adapt through the indirect effect of rerouting to different nodes. However, adaptation is much slower than in the case of the opening of new channels.

To prevent this phenomenon, the conventional method is to introduce a bias D into the estimate:

$$\tilde{T}_i^{n+1}(A_j,B_l) = \tilde{T}_i^n(A_j,B_l) + \alpha(\Delta H - \tilde{T}_i^n(A_j,B_l)) + D \qquad (2.29)$$

The estimate $\tilde{T}_i^n(A_j,B_l)$ is updated regularly, even if A_i does not receive any more packets from A_j. In this case, D is simply added to the current value to give the new estimate. When the path to A_j passing through B_l is cut, the corresponding estimate $\tilde{T}_i(A_j,B_l)$ thus regularly increases, finally ensuring cancellation of the corresponding path. Using the constant D in the estimate also permits reduction of loopback phenomena which are characterized by the fact that packets sometimes return to a node through which they have already passed.

The loopback phenomenon can also be overcome by noting the number of nodes passed through by packets. In this case, ΔH becomes a simple counter which is incremented each time the packet passes through a node and the network discards every packet having a ΔH greater than a certain limit.

2.4.6 Distributed routing

The distributed routing technique [2.13]–[2.14] is a local method in which neighbouring stations exchange messages on the state of the traffic and the network in order to update their delay and routing tables. This method will be described here by assuming that the distance criterion used to optimize the routing is the network transfer time; it is understood that any other criterion, such as the number of nodes passed through by the packets could be used.

At any time, each node such as A_i knows the transfer time $\tilde{T}_i(A_j, B_l)$ of packets between itself and each of the adjacent nodes such as A_j. In effect, A_i knows the queue length on the output line B_l together with the data rate on this line. At regular intervals, each node sends to its neighbours an extract from its table of delays which indicates the smallest estimated transfer time to each of the network nodes. Thus, if node A_j has three output lines B_1, B_4 and B_7, its best estimate of the transfer time to node A_1 is given by

$$\tilde{T}_j(A_1)|_{min} = \text{minimum}\,[\tilde{T}_j(A_1, B_1),\ \tilde{T}_j(A_1, B_4),\ \tilde{T}_j(A_1, B_7)] \qquad (2.30)$$

As soon as $\tilde{T}_j(A_1)|_{min}$ is received, node A_i can update the estimate of the transfer time, t_i, A_1 via B_l in its table of delays by adding the delay on link B_l. The new estimate thus becomes

$$\tilde{T}_i(A_1, B_l) = \tilde{T}_j(A_1)|_{min} + \tilde{T}_i(A_j, B_l) + D \qquad (2.31)$$

where D is a bias whose role was discussed in the previous section and which must be adjusted once and for all in order to stabilize the system and to minimize the mean transfer time.

To describe the operation of distributed adaptive routing, consider a simple network containing five nodes A_1–A_5 in which the adjacent nodes A_2 and A_4 are connected by link B_3 in the A_2A_4 direction and by link B_5 in the A_4A_2 direction. The tables of estimated delays $\tilde{T}_i(A_j, B_l)$ for A_2 and A_4 are given in Table 2.4. The transfer times $\tilde{T}_i(A_j, B_l)$ between A_2 and A_4 are 0.2 on link B_3 and 0.3 on link B_5. (The transfer times in each direction are generally different, since even when the link is symmetrical, the traffic is generally asymmetric which leads to different waiting times in the queues.) Under these conditions, assume that A_2 and A_4 decide to send each other their best estimates of the transfer times. A_4 thus transmits a message to A_2 indicating that its shortest transfer times to A_1, A_3 and A_5 are 0.2, 0.1 and 0.2 respectively; A_2 indicates to A_4 that its shortest transfer times to A_1, A_3 and A_5 are 0.1, 0.2 and 0.3 respectively (Figure 2.9). After this exchange, and

Table 2.4 Distributive adaptive routing. Initial state of delay tables at nodes A_2 and A_4.

Destination nodes	Output lines of node A_2			Destination nodes	Output lines of node A_4		
	B_1	B_3	B_7		B_5	B_6	B_9
A_1	0.1	0.6	0.4	A_1	0.5	0.7	0.2
A_2	—	—	—	A_2	0.3	0.9	0.9
A_3	0.5	0.3	0.2	A_3	0.1	0.9	0.4
A_4	0.9	0.2	0.8	A_4	—	—	—
A_5	0.7	0.9	0.3	A_5	0.3	0.2	0.4

Table of delays of node A_2 Table of delays of node A_4

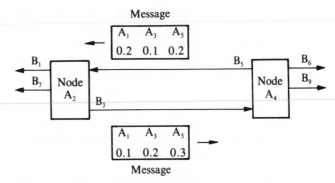

Figure 2.9 Distributive adaptive routing. Exchange of messages between nodes A_2 and A_4.

if the bias D is equal to 0.1, the new estimated transfer times from A_2 to A_1, A_3 and A_5 are obtained by adding the bias 0.1 and the transfer times on link B_3 (0.3) to 0.2, 0.1 and 0.2 respectively; this gives 0.5, 0.4 and 0.5 for column B_3 of the table of delays for A_2 (Table 2.5). Similarly, the new estimates of the transfer times to A_1, A_3 and A_5 from A_4 on B_5 become 0.5, 0.6 and 0.7 respectively.

It can be seen that this adaptive system can, with a certain delay, establish routing tables which minimize the transfer time. The stability and loopback problems are similar to those which are observed with the local routing methods, which justifies the introduction of the bias D. Loopback can be eliminated by including in the method an algorithm which uses, implicitly or explicitly, minimum path trees to prevent a packet passing twice through the same node [2.15].

2.4.7 Hierarchical routing

With non-random routing methods, each node must use a routing table which contains as many lines as there are nodes in the network and a number of columns

Table 2.5 Adaptive routing. State of delay tables at nodes A_2 and A_4 after an exchange of messages. $D = 0.1$.

Destination nodes	Output lines of node A_2			Destination nodes	Output lines of node A_4		
	B_1	B_3	B_7		B_5	B_6	B_9
A_1	*0.1*	0.5	0.4	A_1	0.5	0.7	*0.2*
A_2	—	—	—	A_2	0.3	0.9	0.9
A_3	0.5	0.4	*0.2*	A_3	0.6	0.9	*0.4*
A_4	0.9	0.2	0.8	A_4	—	—	—
A_5	0.7	0.5	*0.3*	A_5	0.7	*0.2*	0.4

Table of delays of node A_2 Table of delays of node A_4

equal to its number of output lines. Also, when routing is adaptive, the nodes must periodically exchange messages for the purpose of updating the routing tables. The size of the routing tables and the amount of the service traffic thus increases rapidly with the number of nodes and becomes unacceptable when the network contains more than a few tens of nodes. To solve this problem, it is necessary to make use of hierarchical organization of the network which groups the nodes into clusters interconnected by a higher-level network. This technique amounts to organizing the network into a hierarchy of sub-networks in a manner analogous to that of conventional telephone networks, where various local networks are interconnected to form a regional network and the set of regional networks is interconnected to form a national network.

An example of a two-level hierarchical network is represented in Figure 2.10. The nodes are grouped here into three areas and all communication from one area to another must pass through the transit nodes. It can be seen that this method greatly reduces the size of the routing tables. These now take account of the nodes of one area only since every packet addressed to a node of another area must necessarily pass through the node which provides the link with the distant area. The price to pay for this simplification of routing is clearly less

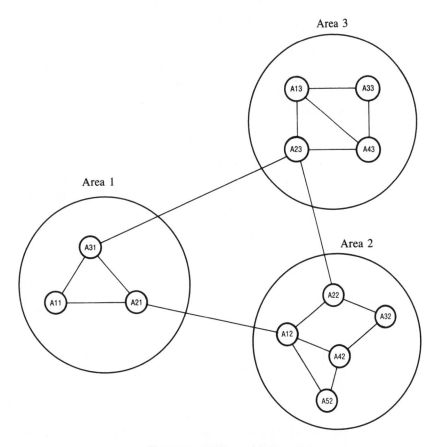

Figure 2.10 Hierarchical routing

satisfactory optimization of the path followed by the packets since these must pass through the prescribed passage points.

The problem of optimizing the number of hierarchical levels has been examined by Kleinrock [2.16] who has shown that routing tables reach a minimum size of $e\mathrm{Ln}(N)$ for a number of hierarchical levels equal to $\mathrm{Ln}(N)$ where $e = 2.718$ and Ln is the natural logarithm. This study also showed that the transfer time is not much longer than with a conventional network when the number of levels is equal to the optimum $\mathrm{Ln}(N)$.

2.5 CONGESTION CONTROL

2.5.1 Traffic jams in networks

Communication networks have only limited resources available to route packets. These limitations occur mainly in the maximum capacity of lines, the processing capacity of the nodes and the size of the buffers which queue the packets in the nodes. When the offered traffic increases, the network suffers **congestion** phenomena, similar to those which are observed with road traffic, and which may lead to a collapse of the network performance.

The network performance can be characterized by various criteria such as the network throughput, the mean packet transfer time and the number of packets lost or in error. If the throughput is measured as the number of packets delivered to users per unit time, an ideal network should be capable of delivering packets in direct proportion to the number of packets entering the network until the maximum network throughput is reached (Figure 2.11(a)). Beyond this limit, the network should be capable of operating at maximum throughput regardless of the traffic offered. In practice, operation of the network deviates from the ideal for a number of reasons which all relate to an inefficient allocation of resources in the case of overload. In particular, the length of the queues for the lines increases rapidly with traffic and exceeds the capacity of the node buffers as

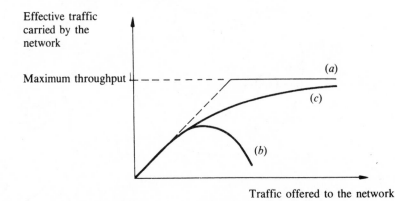

Figure 2.11 Congestion phenomenon: (a) ideal behaviour, (b) congestion, (c) performance of a network with congestion control

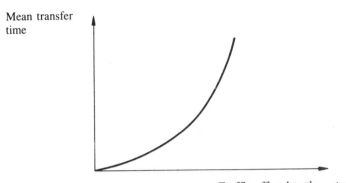

Figure 2.12 Influence of congestion on mean transfer time

maximum capacity is approached. Packets which cannot be stored in memory must be discarded; this causes their retransmission and thus propagates congestion towards the links located upstream by generating dummy internal traffic. There is thus an avalanche effect due to the fact that, above a certain limit, the useful transport capacity of the network decreases when the traffic offered increases (Figure 2.11(b)).

The throughput slackening is only one of the consequences of congestion. All the parameters of the network collapse as soon as congestion occurs, in particular the mean transfer time (Figure 2.12). At the limit, when congestion becomes particularly severe, parts of the network become *deadlocked* due to lack of resources which are themselves used by other parts of the network.

For operation to be satisfactory, the network must be designed to prevent congestion. The simplest solution is clearly to overdesign the equipment in order to operate it far from congestion. This solution is not generally adopted for obvious cost reasons and it is therefore necessary to make use of a number of *preventative* measures [2.17]–[2.21] intended to ensure operation close to the ideal for heavy traffic, as indicated in Figure 2.11(c). The main methods used to prevent congestion are ***traffic control*** which consists of globally limiting the number of packets circulating in the network and ***traffic distribution*** which consists of distributing the traffic equally among the channels. Prevention of congestion is also sometimes achieved by control of packet rate between subscribers that is by *end-to-end flow control*. This type of control, which is intended to match the rate of transmission of the source to the receive capacity of the destination, is incorrectly called congestion control and achieves traffic control and distribution in the network only indirectly. This method is thus questionable, but it is simple to implement so that it has been adopted for several major networks such as ARPANET.

2.5.2 Congestion control by flow control

All congestion control techniques are based on the idea that it is necessary to limit the number of packets in transit in the network in order to prevent network overload. This result can obviously be achieved by end-to-end flow control which limits the maximum number of packets in transit between two users. With such an approach and in a similar manner to that which is used with line procedures, the source marks the packets which it transmits in sequence with a number $P(S)$ defined modulo M. The source cannot send more than W consecutive packets before it receives a packet carrying the number $P(R)$ which acknowledges at least the first of the W packets sent. This *sliding window* procedure ensures that there are never more than W packets in transit between the source and the destination. Consequently, if at a given time there are NP pairs of active correspondents, the procedure ensures that there will not be more than $W \times NP$ packets in transit in the network.

Controlling congestion by controlling the data rate between subscribers has numerous disadvantages related to the fact that it does not operate at the network level. It can be seen immediately that the maximum throughput is limited by the packet transit delay, since the rate of transmission depends on the delay in receiving acknowledgements. Furthermore, if the network contains bottlenecks, it is very possible to have local overload at these points even for a small total number of packets in transit in the network. This leads to operating the system very much below its maximum throughput and hence to an inefficient use of resources.

2.5.3 Congestion control by pre-allocation of resources

In a virtual circuit network, a path is established between the two subscribers at the start of communication by means of a calling packet. This approach can readily be modified by requiring the calling packet not only to establish the route but also to reserve, at each node, the resources which will be allocated to the communication being established. In particular, buffers can be assigned in each node and for each active virtual circut; the size of these corresponds at least to the maximum number of packets in transit, that is the window size. The resource pre-allocation method prevents congestion since the procedure rejects any call for which resources cannot be allocated. This technique is thus quite similar to that which is used in conventional telephone switching where the calling procedure establishes a physical path between the subscribers and activates the corresponding resources for the duration of the call. In the two cases, the resources are very badly used, since they are not shared between users when these are inactive. Pre-allocation of resources thus leads to overdesign of the network.

2.5.4 Isarithmic control

The **isarithmic control method** [2.22] limits directly the number of packets in transit on the network by means of a system of tokens. This technique is thus a priori more efficient than indirect approaches based on end-to-end flow control.

With this method, M *tokens* (credits) flow continuously on the network and a packet can be transferred only if it is in posession of a token. Thus it is impossible for the number of packets in transit on the network to exceed M. The value of M is clearly chosen so as to avoid congestion while ensuring reasonable performance. In practice, each node stores several tokens and sends surplus tokens to neighbouring nodes so as to ensure equal sharing of tokens throughout the network. A packet can enter the network only if the input node has at least one token; if not, the packet is rejected. If the packet is accepted, it seizes the token and returns it on leaving the network.

Various simulations [2.23] have shown that when M is suitably chosen, isarithmic control reduces the network throughput only very slightly and enables the mean packet transfer time to be very significantly reduced when the network is heavily loaded (Figure 2.13). Isarithmic control thus has useful characteristics in respect of network performance.

Practical realization of isarithmic control raises difficult problems of circulation and protection of tokens. To prevent distribution of tokens causing substantial additional traffic, they can generally be incorporated in data packets. This ensures that the service traffic due to tokens remains small. In contrast, the policy to be adopted to distribute the tokens in the network in an optimum manner is not clear. If the tokens are distributed uniformly, each station can expect to gain easy access to the network even if the traffic is high. However, this approach limits the available number of tokens at a given time at each node and this introduces long delays for users with heavy traffic since the node to which they are connected must be continuously begging tokens from its neighbours. The isarithmic method thus provides global control which does not operate effectively at the points where there is a heavy concentration of traffic.

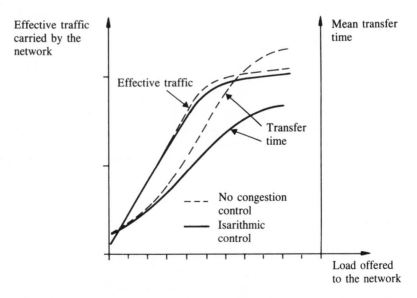

Figure 2.13 Performance of a network with and without isarithmic control

Notice finally that maintaining the integrity of the tokens poses serious problems. If many tokens are lost for any reason, the capacity of the network is correspondingly reduced and this is unacceptable. It is, therefore, necessary to provide a control and recovery procedure for tokens; realization of this is difficult since there is no simple means of knowing the total number of tokens in circulation on the network at a given time. Because of these difficulties, the isarithmic method has not so far been realized practically.

2.5.5 Congestion control by link load control

The methods of congestion control presented above do not provide any means of preventing local congestion in the parts of the network where there is heavy traffic. To remedy this situation, it is possible to provide more refined congestion control based on the effective rate of use of the various lines.

An approach of this type will be described here in the context of the CIGALE network [2.24]. With this method, the activity on each line is represented by a load index LI which varies between 0 and 1 and is such that LI takes the value 0 for a free line and the value 1 for an overloaded line. The value of LI is updated in an adaptive manner from observing the busy or free state of the line at regular intervals. Hence, denoting the state, busy or free, of the line by the value 1 or 0 of the binary variable f, the new value LI^n of the load index is established from the previous value LI^{n-1} by the equation

$$LI^n = \alpha LI^{n-1} + (1 - \alpha)f \qquad (2.32)$$

where α is a constant less than 1 which determines the rate of adaptation. When the load index of the line exceeds a limiting value LI_1, for example 0.7, the node located at the output of the line checks the source address of the packet which has just arrived on this line and sends a warning message to the source of the packet to indicate that its rate of transmission should be reduced. The source node then reduces its traffic for a predetermined period after which it returns to its normal rate until it again receives a warning packet. It can be seen that the system does not introduce any control when the network is lightly loaded, and this avoids artificially limiting the rate of packets when these can be transferred under suitable conditions. In contrast, when congestion appears at certain points of the network, the system adapts selectively by reducing only the traffic directly affected by the congestion. It can be expected that under these conditions the performance of the network with respect to offered traffic will be very close to the ideal.

2.5.6 Recovery from congestion

Most prevention techniques do not guarantee congestion-free operation. In particular, global methods always leave open a risk of local congestion at certain points. It is, therefore, necessary to provide recovery methods which ensure resumption of normal operation when the congestion has settled. Recovery methods can also be used alone without the use of any prevention technique.

The simplest method to eliminate congestion is to discard some of the packets causing the congestion; this clearly tends to re-establish normal network operation. To be effective, such a technique must be used selectively. It would not be advisable to discard service packets, since these serve to control network operation. Similarly, it is unwise to drop acknowledgement packets, since they are used to release certain network resources. Finally, one has to take account of the fact that the network operates a number of different retransmission mechanisms to recover lost packets. There is a risk that, if the network rejects a large number of packets in transit, in order to resolve its congestion problem, this will cause only a temporary relief followed by a serious relapse due to retransmission traffic. To avoid this kind of effect, the best policy is to discard packets as soon as they enter the network rather than waiting until they have already passed through several nodes. Such an approach limits retransmission traffic, to a large extent, to the nodes and lines at the input to the network.

In order to reduce the maximum number of packets which must be discarded to reduce congestion, it is appropriate to adopt techniques which enable the network resources to be put to the best use. In practice, the resource which fails first when congestion occurs is generally the available memory at the various nodes. In a typical node (Figure 2.14), the packets arriving on the input lines are stored in an arrival queue. These packets are then processed by the node processor which routes them to the various output lines according to their destination. As the capacity of the output lines is limited, the packets to be transmitted on them must be stored in queues whose length increases very rapidly when the traffic flow approaches the line capacity. If the size of the queues exceeds the available memory capacity, the node can only discard the excess packets. The question which arises is finding the best way of managing the node memory in order to limit the number of discarded packets.

An obvious first strategy is to share the node memory without restriction among all the queues. With such an approach, if M is the size of the buffer memory associated with each output line and K is the number of buffers available at the node, all the buffers can be allocated to the line which carries the heaviest traffic, for example S_1. Then $M = K$ and the packets addressed to other lines such as

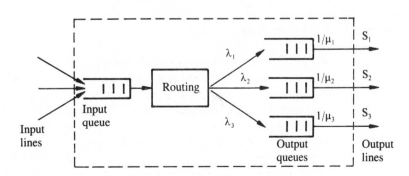

Figure 2.14 Packet switch

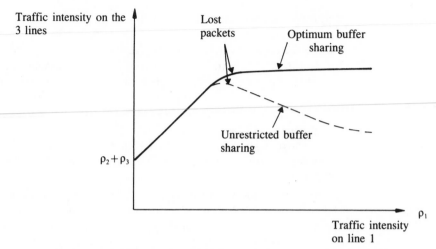

Figure 2.15 Effect of a limitation of buffer sharing on packet loss

S_2 and S_3 must be discarded due to the lack of free buffers. It can be shown [2.25] that this method is inefficient, since increasing the memory size corresponding to the queue on the most heavily loaded line S_1 does not increase its throughput, which is already maximum, while the lack of memory on the other lines reduces their traffic handling capacity (Figure 2.15).

Equal sharing of the memory among all the output lines could also be considered, hence $M = K/L$ where L is the number of output lines. This technique gives even worse results than the previous one since it does not permit the node to adapt the memory resources to the traffic requirements. In practice, the best results are obtained with a heuristic method which consists of sharing the memory between the output lines as a function of the traffic; but the maximum size of the buffer memories on each line is limited in order to retain some available buffers for the lines with low traffic. This technique, which is illustrated in Figure 2.15, prevents discarding of packets on low traffic lines while only slightly reducing the throughput on the heavy traffic lines, since these already operate at their maximum rate.

2.5.7 Deadlock and deprivation

It was shown above that a network overload leads to a collapse of the performance which relates to inefficient utilization of the network resources. In extreme cases, the performance degradation can result in a total blockage of some elements of the network. This situation is called *deadlock* and corresponds to the case where two or more processes are blocked while waiting for resources which are already allocated to other processes.

In its simplest form [2.26]–[2.27] (*direct store and forward deadlock*), deadlock involves two adjacent nodes A_1 and A_2 which are in a situation such that all the buffers of node A_1 are allocated to line A_1A_2 while all the buffers of node A_2

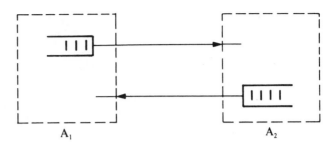

Figure 2.16 Simple deadlock

are allocated to line A_2A_1 (Figure 2.16). Under these conditions, there are no more available buffers in A_2 to receive acknowledgement packets from A_1, and the buffers which are in the output queue cannot be freed for this purpose, since A_1 is unable to receive packets from A_2 due to the lack of available buffers. The system thus remains blocked indefinitely. It is relatively easy to prevent this type of simple deadlock by dividing the available buffers at each node into two groups of which one is reserved for output lines and the other is reserved for input lines. However, the problem becomes much more complex when the deadlock involves several nodes which mutually prohibit access to buffers which they require to pursue their activity. In this case, it is generally necessary to make use of a prevention method based on sharing of resources in *ordered classes* of which the principle is described below.

It has been seen that deadlock involving two nodes can be prevented by dividing the buffers at each node into two classes corresponding to the input and output lines of the node. The method can be generalized to a network of N nodes in which the longest path between a source and its destination passes through G nodes. The buffers at each node are divided into $G + 1$ classes numbered from 0 to G and a buffer of number i of a node A_j can send packets only to a buffer of number $i + 1$ of an adjacent node A_{j+1}. With this approach, a packet can enter the network only if there is a free buffer of class 0 in the source node. As it propagates through the network, the packet passes into buffers whose class number increases by 1 each time the packet passes through a node. If the packet reaches its destination before being in class G, the corresponding buffer is freed and this permits a buffer of an immediately lower class to transmit a packet and become free in turn. If the packet reaches class G, the buffer is also freed if the node which it is in corresponds to the destination of the packet. Otherwise, the packet is discarded since it must have followed a prohibited route, as the maximum path length in the network is equal to G nodes. In all cases, the class G buffer is freed and this permits the system to continue operating by preventing all risk of blocking.

There are numerous other forms of deadlock in networks. Among those which will be presented briefly here are: deprivation, reassembly deadlock, piggyback deadlock and user propagation deadlock.

Deprivation can appear in a network where the packets are assigned different priorities, for example priority 1 for dialogue packets and priority 2 for normal packets. It is clear that if the traffic of priority 1 packets is too large, transmission of priority 2 packets can be delayed indefinitely. This phenomenon is not too serious when high priority traffic is limited. However, segregating packets into priority levels can have very serious consequences in relation to deadlock, even if the high priority traffic remains small. If all the buffers of a node A_i are full with packets of priority 2 which are awaiting acknowledgement, the latter cannot reach A_i if the nodes adjacent to A_i contain priority level 1 packets. Hence blocking occurs. This type of situation can be prevented by applying the priority rules in a flexible manner or providing separate buffers for high priority packets.

Reassembly deadlock can appear when messages from the user are segmented into packets to be carried more efficiently by the network. These packets must be reassembled into messages by the destination node before delivery to the recipient. As the packets do not necessarily arrive at the destination node in the order in which they were sent, the destination node must store the received packets in its buffer memory until it has complete messages which can be delivered to the user. If the memory is completely filled with incomplete messages, the destination node can neither receive packets to complete the messages nor deliver messages to the user; this causes a deadlock. This type of impairment can be avoided by reserving the memory space required to store at least one complete message. Reassembly deadlock of a second order can appear when all the memory of the destination node A_i is full or reserved for messages being reassembled. If nodes A_{ij} adjacent to A_i are themselves full to maximum capacity with packets destined for A_i but not corresponding to the messages being reassembled by A_i, these packets cannot be accepted by A_i and the system becomes blocked. One simple solution is to reserve a buffer on each input line of A_i. Node A_i thus receives packets from the adjacent nodes A_{ij}. It discards the packets for which there is no space for reassembly in its memories until it receives the packets which enable it to complete a message and send it to the destination thereby releasing the corresponding memory.

Piggyback deadlock can occur when acknowledgements are inserted into text packets destined for a node A_i. It can happen that A_i cannot receive more packets due to a lack of buffers which can only be freed after receipt of the acknowledgements. As the latter are piggybacked into text packets, they cannot be received by A_i and blocking occurs. It is possible to prevent piggyback deadlock by requiring acknowledgements to be sent separately after a specified delay.

The network can also suffer from user deadlock which propagates if no particular care is taken. If users do not use deadlock prevention mechanisms, they can very well be in a situation of the type represented in Figure 2.16, where all their buffers are taken up by output lines. In this case, the users are blocked as a consequence of the lack of input buffers; this causes a pile-up of packets which are in transit on the network and cannot be delivered. The deadlock thus tends to creep step by step to other nodes of the network. This type of deadlock is automatically prevented by end-to-end flow control mechanisms which limit the number of packets in transit for each source/destination pair.

2.6 X.25 PROTOCOL

2.6.1 Overview

To satisfy the rapidly increasing requirements for data transmission, most industrialized countries have installed public packet switching teleprocessing networks. This has led the CCITT to define a standard interface with this type of network in order to avoid the development of mutually incompatible systems. The result of this effort is the X.25 protocol [2.28]–[2.33] which is used with most public packet switching networks.

The X.25 protocol specifies the rules for communication between the user, called the *data terminal equipment (DTE)*, and the public network input station, called the *data circuit terminating equipment (DCE)* (Figure 2.17). In practice, the electrical interface between the user and the network is located between the DTE and a DCE located with the user which generally reduces to a single modem. The DCE is connected to the nearest network node by a transmission line and the intelligent part of the protocol is handled by the DTE and the *data switching equipment (DSE)* of the node to which the DCE is connected. In the following part of this section, and to conform to Recommendation X.25, it will

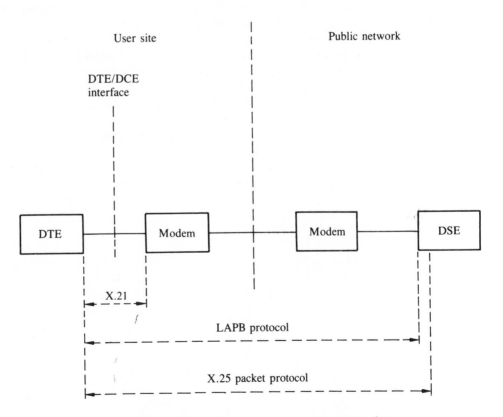

Figure 2.17 X.25 protocol. Network/user interface

be assumed that the X.25 protocol defines the exchanges between a DTE and the corresponding network DCE, although this approach can lead to confusion (Figure 2.18).

The X.25 protocol does not define the internal operation of the network, that is the manner in which the packets are routed between the various DSEs corresponding to the nodes of the network. Practical implementation and performance can thus vary appreciably from one network to another. In contrast, X.25 specifies very precisely the interface between the network and the user in such a way as to permit all X.25 DTEs to be connected to any public network which conforms to the protocol. Recommendation X.25 defines three protocol levels which correspond approximately to the first three layers of the OSI model (Figure 2.18).

Level 1 of X.25 specifies the physical, electrical and logical characteristics of the interface between the DTE and the DCE. This level, which corresponds to the transmission of bits by a physical channel, is specified in detail in CCITT Recommendation X.21 (Vol. I, Chapter 2). This recommendation relates to access to a digital network. As networks of this type are rarely available and the same applies to the corresponding DTE interfaces, access to packet switching X.25 networks is generally achieved by way of the telephone network using modems equipped with a V.24/RS.232.C interface. In this case, the interface is specified by CCITT Recommendation X.21b with synchronous transmission.

Level 2 of X.25 defines the point-to-point transfer of frames between the DTE and the DSE. It is, therefore, a line procedure for which Recommendation X.25 specifies two variants which are the *link access procedure (LAP)* and the *balanced link access procedure (LAPB)*. These line procedures are both bit oriented and are very similar to the HDLC procedure (Vol. I, Chapter 4); in particular there is compatibility between LAPB and the ABM mode of HDLC.

The third level of the X.25 protocol concerns the structure of text and control information in packets. This ***packet level protocol (PLP)*** defines the format of the data field of the frames exchanged at level 2. Hence, the text transmitted by the user is first augmented at level 3 with a packet header which specifies, for example, the logical channel number or the type of packet (Figure 2.19). The

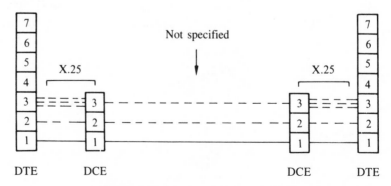

Figure 2.18 Layers of X.25 protocol

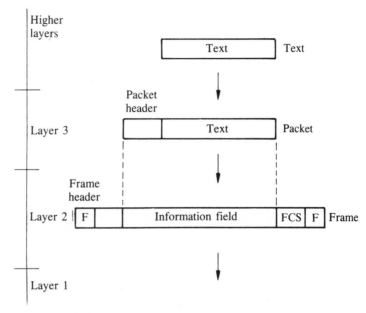

Figure 2.19 Format of various levels of X.25 procedure

packet is then sent to level 2 which incorporates it into the information field of a line frame. This frame is then transferred to the network by the physical layer at level 1. Level 3 of the X.25 protocol corresponds approximately to the network layer (also level 3) of the OSI model. It will be seen, however, that this correspondence is only approximate and in practice X.25 permits partial or total end-to-end control to be provided in accordance with the options chosen and the level of confidence which the user assigns to the network. The X.25 protocol thus exceeds level 3 in order to support all or part of the functions normally provided by the transport layer (level 4) of the OSI model.

The X.25 protocol was developed initially to provide a permanent or switched virtual circuit service. The physical connection between the network and the user can also be permanent or switched. In order to ensure optimum utilization of the user/network link on the line connecting the DCE or DSE, the protocol provides a multiplexing facility by offering several *logical channels* represented by a logical channel number which must be inserted into all packets. This enables several virtual circuits to be multiplexed on a single real access circuit to the network.

The X.25 protocol was first standardized in 1980 [2.28]. It subsequently became necessary to make modifications to ensure compatibility between this protocol and the OSI model and to specify some points omitted from the standard. This led to the introduction in 1984 of a new version of the X.25 protocol [2.29]. The 1980 X.25 protocol provided an optional *datagram* service together with a similar service called **fast select** which is based on transfer of user data at the time of setting up the virtual circuit, with immediate clearing of the circuit at the end of the set-up phase. The 1984 X.25 protocol retained only the fast select service which was more compatible with the virtual circuit service than the datagram service.

The X.25 protocol is for users who have equipment which operates in synchronous mode and implements the protocol at the X.25 packet level. Equipment of this type, called DTE-P, is relatively expensive and it was necessary to offer the facility of connecting simple terminals operating in character mode to the public packet switching network; these are here called DTE-C. This can be realized by means of auxiliary *packet assembly disassembly (PAD)* equipment which manages the network protocol functions and provides conversion between synchronous and asynchronous line procedures [2.28]–[2.29]. As PADs are relatively costly, they are generally shared among several terminals which leads to their use as *cluster controllers*. In practice, the PADs can be part of the public network or can be located with the user. The result is a configuration such as that of Figure 2.20 where a DTE-P operating in packet mode with synchronous transmission is connected to asynchronous DTE-C terminals by the PADs; these can be part of the network or can be located with the user.

The packet assembly and disassembly service is specified by CCITT Recommendation X.3. Recommendations X.28 and X.29 define the protocols for exchanges between a PAD and a DTE-C or a DTE-P respectively.

In some cases, it can be useful to provide a packet assembly and disassembly service for synchronous terminals which do not have the X.25 packet functions. This service has not been standardized, but some manufacturers provide special PADs which are designed for this application [2.34].

A public packet switching network is available in France with the name TRANSPAC and in Switzerland with the name TELEPAC.

In practice, use of the same word 'network' to designate the level 3 protocol of the OSI model and the communication network which connects the users is a source of confusion. This has led to the ISO recommending the name 'sub-network' to designate a communication network in the conventional sense of the term. With this terminology, a public communication network must be called a 'sub-network'. The word 'sub-network' is itself a source of confusion since it tends to imply membership of a larger network. It will be used here only in exceptional cases.

2.6.2 LAP and LAPB line procedures

The LAP and LAPB line procedures form level 2 of the X.25 protocol. These two synchronous procedures are bit oriented and directly derived from the HDLC (Vol. I, Chapter 4); they have, in particular, the same frame type and structure. They are normally used for network access with point-to-point operation on a single physical circuit in full duplex [2.29].

The LAP and LAPB *single link procedures (SLP)* provide basic access to public packet switching networks and are generally used as such. However, it is sometimes necessary to connect several parallel circuits to the network in such a way as to prevent system interruption in the case of failure of one of the circuits. To meet this requirement, the CCITT has provided a ***multi-link procedure (MLP)***

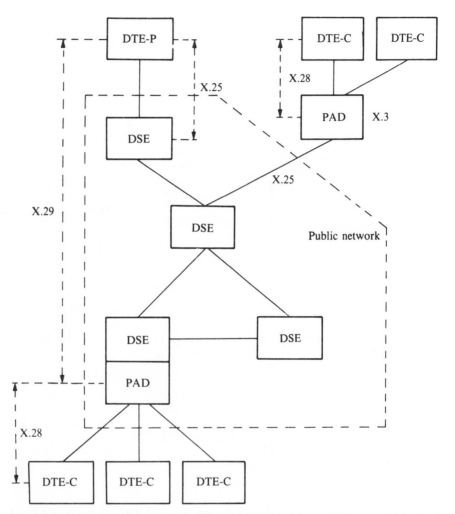

Figure 2.20 Use of a packet assembler/disassembler for access to a public packet switching network

which permits sharing of packet transmission on the various single link procedures used between the DTE and the DCE. In this case, the single link procedures can only be of the LAPB type.

The LAP procedure operates in *asynchronous response mode (ARM)* on an unbalanced line where each station incorporates a primary text transmission function and a secondary text reception function (Vol. I, Figure 4.10). The LAP procedure is very similar to the HDLC procedure in the asynchronous response mode, but it is incompatible with it. It uses I, RR, RNR and REJ information and supervisory frames with modulo 8 sequence numbering and unnumbered SARM, DISC, CMDR and UA frames. As the configuration is symmetrical, the bidirectional link can be considered to consist of two unidirectional links in opposite directions. This requires the two half links to be established individually

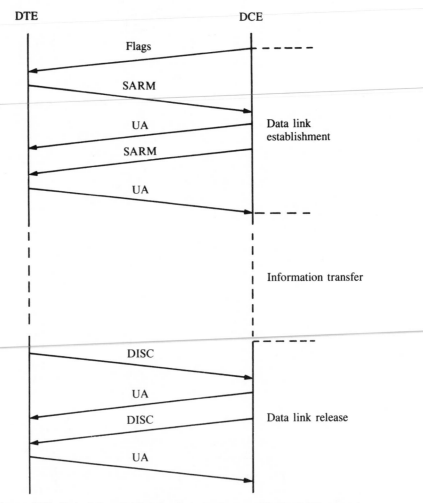

Figure 2.21 Data link establishment and release with the LAP procedure

by SARM frames which are acknowledged by unnumbered UA frames (Figure 2.21). Before these operations, the DCE indicates that it is ready to establish the link by sending successive flags. Clearing of the link is performed similarly by two DISC command frames.

The LAP procedure has the disadvantage of incompatibility with HDLC. Because of this, it is being replaced by the LAPB procedure which is compatible with the HDLC procedure in asynchronous balanced mode (ABM) with options 2.8 or 2.8 and 10. The DTE and DCE are combined stations in this case (Vol. I, Figure 4.11) of such a kind that the link is established by a single initialization frame SABM (Figure 2.22). In a similar manner, the link is cleared by a single disconnection frame DISC. Information transfer is performed according to the principles given in Chapter 4 (Vol. I) using information frames I and supervisory

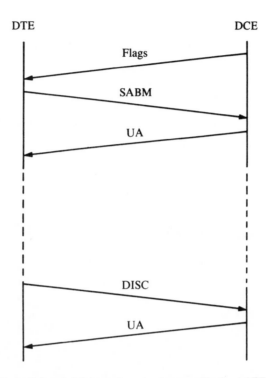

Figure 2.22 Data link establishment and release with the LAPB procedure

frames RR, RNR and REJ which are numbered modulo 8 or, optionally, modulo 128. As the link connects two points, there can be no ambiguity concerning the identity of the source and destination of a frame. In contrast, as the stations are combined, each can transmit commands and replies equally well. Two distinct addresses A and B are therefore used to distinguish commands and responses. Address A, which has the value 11000000, indicates commands transmitted by the DCE and the responses received by it. Similarly, address B (10000000) corresponds to commands sent and responses received by the DTE.

The *multilink procedure* makes use of several parallel data links of the LAPB type on which it distributes frame transfer. This mode of operation introduces new problems since frames no longer necessarily arrive sequentially at the destination. More generally, the multilink procedure must implement some functions to control the group of links, for example reset of the group or to put one of the links out of service. To achieve this, the multilink procedure uses frames which have the same general structure as LAPB frames with the exception of an additional two byte command field (Figure 2.23). This field contains, among others, a 12-bit number which is used to maintain frame sequence. In order to distinguish between multilink and single link operation, multilink frames are identified by two addresses C and D which are different from addresses A and B of LAPB frames.

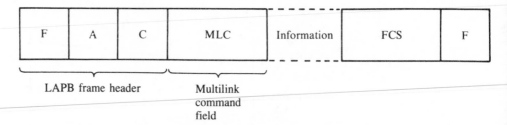

Figure 2.23 Format of a multilink frame

2.6.3 X.25 packet level protocol

The *X.25 packet level protocol (PLP)* provides a *switched virtual circuit service* on a public packet switching network. The switched virtual circuit establishes temporary communication between two network users, with calling and clearing procedures which enable any two subscribers on the network to be connected at any time. With this type of service, the packet switching network plays a similar role to a circuit switching network such as the conventional telephone network. For applications which require a permanent connection between two stations, X.25 networks have the facility of establishing, on subscription, *permanent virtual circuits* whose function is similar to that of leased lines in conventional networks. Finally, users have the option of using the X.25 network with a *fast select service* which can be considered as a special type of *datagram service*.

The X.25 protocol was originally designed for public packet switching networks. It is now universally accepted and a large amount of equipment today incorporates the X.25 protocol functions. Because of this, usage of the X.25 protocol tends to extend to applications which were not originally foreseen. Hence, for example, some organizations establish private X.25 networks using leased lines. Similarly, some local area networks use the X.25 packet level protocol as the network layer [2.35]. These approaches are very useful in some cases, since they permit the cost of interfaces with the communication controller to be reduced by means of the multiplexing facilities provided by X.25.

The X.25 packet level protocol provides the network layer functions of the OSI model and, if necessary, some of the transport layer functions under conditions which will be specified in Section 2.7. At this level, communication is ensured by transferring packets of one of the types indicated in Table 2.6. Establishment of a virtual circuit between two DTEs is performed by exchanging CALL REQUEST, INCOMING CALL, CALL ACCEPTED and CALL CONNECTED packets. These packets contain the network addresses of the called and calling subscribers. After these operations, the virtual circuit is established and the data packets can be exchanged without any address reference since they all follow the path defined by the virtual circuit. Actually, the situation is slightly different, since the physical circuit and the data link between the DTE and the DCE are multiplexed so that the same DTE can operate with several virtual circuits simultaneously. This requires each virtual circuit operated on the same link to be identified by a distinct *logical channel number* which remains

Table 2.6 X.25 packet types.

Type		Service	
From DCE to DTE	From DTE to DCE	Switched virtual circuit	Permanent virtual circuit
Call set-up and clearing of virtual circuits			
INCOMING CALL	CALL REQUEST	x	
CALL CONNECTED	CALL ACCEPTED	x	
CLEAR INDICATION	CLEAR REQUEST	x	
CLEAR CONFIRMATION	CLEAR CONFIRMATION	x	
Data and Interrupts			
DATA from the DCE	DATA from the DTE	x	x
INTERRUPT by the DCE	INTERRUPT by the DTE	x	x
INTERRUPT CONFIRMATION	INTERRUPT CONFIRMATION	x	x
Flow control and reset			
RR from the DCE	RR from the DTE	x	x
RNR from the DCE	RNR from the DTE	x	x
	REJ from the DTE	x	x
RESET INDICATION	RESET REQUEST	x	x
RESET CONFIRMATION	RESET CONFIRMATION	x	x
Restart			
RESTART INDICATION	RESTART REQUEST	x	x
RESTART CONFIRMATION	RESTART CONFIRMATION	x	x
Diagnostic			
DIAGNOSTIC		x	x
Recording			
RECORD CONFIRMATION		x	x
	RECORD REQUEST	x	x

valid for the lifetime of the virtual circuit. Assignment of a number to a virtual circuit is performed at the time of establishment, by including a logical channel number, chosen from the free ones, in the call request packets. This enables DCEs located at two ends of the network to associate logical channel numbers with the addresses of correspondents, and hence with the path followed by the packets on the virtual circuit. After initialization, all packets must contain the number of the logical channel to which they belong so that the network can provide correct routing. The logical channel numbers clearly have a purely local significance, since they refer to the multiplexed data link between the DTE and the local DCE. The result is that packets exchanged between two correspondents on the same virtual circuit generally have different numbers for each correspondent; this can be a source of confusion for the novice.

Text is transferred on a virtual circuit by DATA PACKETS with a sliding window procedure which is very similar to that used with HDLC (Vol. I, Chapter 4). The principle of packet transmission with anticipation of acknowledgements is used again, with send and receive numbers. The acknowledgements can be piggybacked in the data packets, transmitted in the opposite direction using the receive number, or transmitted by the RECEIVE READY (RR) supervisory packets which contain the receive number.

X.25 packets are not checked by an error detecting code. It is assumed that transmission errors are corrected in the data link layer so that the only errors which require correction at level 3 of X.25 are those which relate to packets lost or received out of sequence following a network fault. This type of defect, which is uncommon, can be detected by checkpointing the packets without an error detecting code. Under these conditions, it can be seen that the mechanism of acknowledging data packets by sequential numbering serves principally to provide flow control on the virtual circuit. Control of the rate of sending data packets is effected by adjusting the rate of sending acknowledgements and using RECEIVE NOT READY (RNR) packets which acknowledge the data packets already received and stop transmission of new packets. Optionally, some X.25 networks provide a facility to request retransmission of packets by means of a REJECT (REJ) packet whose receive number indicates the number of the first data packet which must be retransmitted.

Clearing of the switched virtual circuit is performed in a similar manner to establishment by the exchange of CLEAR packets.

Errors and operational defects are taken care of by RESET, RESTART and DIAGNOSTIC packets. Finally, RECORD packets are used to request recording of supplementary services or to obtain the present values of these services.

2.6.4 X.25 packet formats

All X.25 packets contain at least three bytes which contain a general format *identifier*, a *logical channel identifier* and a *packet type identifier* (Figure 2.24). According to the packet type, the basic header can include extra bytes which contain additional fields. All packets have a length which is an integral number of bytes. As for the HDLC procedure, X.25 packets can be subdivided into *data packets* which contain send and receive numbers, *supervisory packets* which include receive numbers and *unnumbered packets* which are used to control the virtual circuit.

Text is transmitted in the form of DATA packets whose format is represented in Figure 2.25; they combine the user's data with a header consisting of three bytes and containing the following fields:

Q (1 bit): This bit, whose exact significance is not specified by the X.25 protocol, is intended to permit higher levels to specify the significance of the text part of the packet, for example to indicate whether the text corresponds to data or

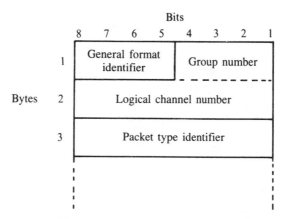

Figure 2.24 Basic header of an X.25 packet

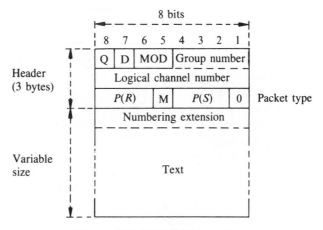

Figure 2.25 X.25 DATA packet

control information. What is considered to be text at one level can very well be control information at a higher level. The X.29 protocol uses the Q bit to indicate the difference between normal packets, for which Q is set to 0, and packets used for commands and responses exchanged with the PAD, for which the Q bit is set to 1.

D (1 bit): The D (*confirmation of delivery*) bit is used to indicate the source of the acknowledgement which can be either the local DCE when D is set to 0 or the destination DTE when D is set to 1. Notice that, in the latter case, routing control is performed end-to-end which corresponds to an X.25 transport protocol operation, on condition that the quality of the network is sufficient for additional error checking or recovery to be not required.

Logical channel identifier (12 bits): The number of the logical channel on which the packet is to be routed is indicated in the 12-bit header field which defines up to 4096 multiplexed logical channels on the user/network physical link. The logical channel identifier is subdivided into a *logical group number* (4 bits) and a *logical* *channel number* (8 bits). The logical channel identifier which can be used are established by mutual agreement at the time of installation. The rule imposed by X.25 is to number all the permanent virtual circuits in increasing order starting from 1 and to reserve the higher numbers, again in increasing order, for switched virtual circuits. All packets sharing the same logical channel must contain the channel identifier in their header, whatever their format. Logical channel identifiers define only the circuit between a user and the network. They are not, in principle, the same on transmission and reception.

MOD (2 bits): The two bits of the MOD (*modulo*) field are set to 01 if the packet sequence numbers $P(R)$ and $P(S)$ are defined modulo 8 and set to 10 when the sequence numbers are defined modulo 128.

$P(R)$ and $P(S)$ counters: $P(R)$ and $P(S)$ are two sequence numbers each containing 3 or 7 bits according to the setting of MOD; they serve the same purpose at the network level as the numbers $N(R)$ and $N(S)$ used by the line procedure. In the case where MOD is equal to 01, $P(S)$ indicates the sequence number of the transmitted packet modulo 8 and $P(R)$ represents the number of the next expected packet. The numbers $P(R)$ and $P(S)$ serve to control the flow between subscribers by means of the following mechanism: The source DTE A increments its $P(S)$ counter each time it sends a packet, and it is permitted to send several data packets before receiving an acknowledgement $P(R)$. To limit the number of packets in transit between the source and destination, the users and the network define, by prior agreement, a window W which specifies the maximum number of packets which can be transmitted before receipt of an acknowledgement. Under these conditions, the last $P(R)$ received indicates to A that all packets numbered less than $P(R)$ have arrived correctly and it can send packets such that $P(R) \leqslant P(S) < P(R) + W$. When D has the value 1, the acknowledgement indicates correct delivery of the packets to the destination DTE. If D has the value 0, the acknowledgement indication serves only for flow control. In this case, the local DCE regulates the rate of sending its acknowledgements in order to control the user's transmission rate, and the acknowledgements do not in any way indicate correct transfer of the packets. When the $P(R)$ and $P(S)$ counters are defined modulo 128, the packet header contains an additional byte to extend the numbering.

M (1 bit): When it is set to 1, the M (*more*) field, *more data*, specifies that the data packet is part of a message. In this case, all the packets of the message are such that M = 1 except the last packet for which M = 0. Only complete packets, that is those which have the maximum size specified by prior agreement can have a continuation indication M = 1. The logical grouping of packets belonging to the same message enables the network to modify the packet size in order to suit the requirements of traffic flow better and possibly to deliver packets to the destination of a size which conforms to requirements. The M bit is part of the third byte of the header which specifies the type of packet. This byte also includes $P(R)$ and $P(S)$ and a control bit which identifies the type of packet and is set to 0 to indicate a data packet.

Data field: The data field has a variable size which is a multiple of a byte. The extent of this field is limited by the maximum size of the packets and is established by prior agreement at 16, 32, 64, 128, 256, 512, 1024, 2048 or 4096 bytes. By default, the maximum size is set at 128 bytes.

The *supervisory packets* RR, RNR and REJ have a length of three bytes, except with the option of extended sequence numbering modulo 128 which requires a one-byte extension (Figure 2.26). They are distinguished from data packets by their packet identifier of which the first bit is set to 1. Supervisory packets are used only to signify the acknowledgement of, or to request retransmission of, data packets. They therefore contain only one receive number field $P(R)$ whose length is three bits when numbering is modulo 8.

Unnumbered packets have variable formats according to their type, but their header always starts with the three basic bytes indicated in Figure 2.24 with the first bit of the third byte set to 1 to distinguish them from data packets. The formats of non-sequential packets will be described in more detail in the following sections.

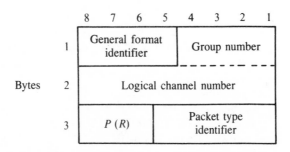

Figure 2.26 Format of supervisory packets RR, RNR and REJ

2.6.5 X.25 virtual switched circuit operation

When a user wishes to establish an X.25 virtual circuit link with another network subscriber, he must first have the use of a physical circuit between his DTE and a network node. If the circuit is permanent, this circuit establishment phase does not exist. In contrast, if the line passes through the switched network, the user must establish the connection by using the X.21 or X.21b call and set-up procedures described in Chapter 2 (Vol. I). When the physical connection is established, it is necessary to initialize the data link by operating the initialization phase of the line procedure. In the case of the LAPB (HDLC–ABM) procedure, this initialization reduces to the DCE sending flags followed by despatch of an SABM frame which the DCE must acknowledge with a UA frame (Figure 2.22). The logical link with the network is now established and the calling DTE can start to establish the virtual circuit with the called DTE by sending a CALL REQUEST packet to the network; the format of this packet is indicated in Figure 2.27. The header of this packet contains the three standard packet bytes to which are added a byte which specifies the size of the source and destination address fields (SALF, DALF) which follow. The header then contains

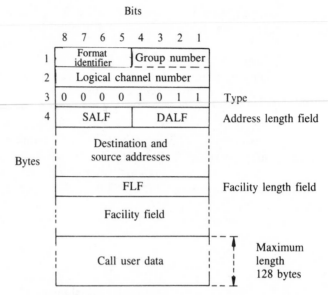

Figure 2.27 X.25 CALL REQUEST and INCOMING CALL packets

a byte which defines the size of the facility field which is itself followed by the variable size field in which the optional facilities requested by the user are specified. These facilities can, for example, be selection of the throughput on the virtual circuit; they will be described in the following section. Finally, the last field of the packet is a user data field containing up to 128 bytes which can be used to transmit a password, for example.

To establish a virtual circuit connection, the calling subscriber A chooses the highest free number from the logical channel numbers which are available and places this number in the corresponding field of the CALL REQUEST packet for which he establishes the other fields as required. This packet must contain the calling and called network addresses in order to permit establishment of a virtual circuit between the two correspondents. The called address is clearly always necessary, but the calling address can be omitted in some cases if the caller is connected to the network by a permanent link. The CALL REQUEST packet is then sent to the local DCE which routes it in turn to the DCE connected to the called subscriber B (Figure 2.28). The DCE connected to B chooses the smallest free number from the available logical channel numbers in the direction of B. It then sends the CALL REQUEST packet to DTE B by simply replacing the transmission logical channel number with the local logical channel number. The CALL REQUEST packet then takes the name INCOMING CALL. If B accepts the call, it returns a CALL ACCEPTED packet which is routed to the DCE adjacent to A where it takes the transmission logical channel number to arrive at A under the name CALL CONNECTED which completes the initialization phase and permits the transfer of DATA packets to start.

Once the virtual circuit is established, the two DTEs can transmit DATA packets using a technique similar to that used with the HDLC procedure; for the REJECT (REJ) packet can request retransmission of all the DATA packets from the sequence number $P(R)$ which it contains.

In practice, the fact that the packets are transmitted on a network and not a single link makes the flow control mechanism more complicated than with line procedures. Consider, firstly, the case corresponding to transfer of DATA packets with confirmation of delivery, that is with the D bit set to 1. The flow control mechanism is thus apparently similar to that of the line procedures, since the packets are sent in anticipation of acknowledgements up to the credit limit allowed by the window W, and renewal of credit due to acknowledgements sent by the destination DTE (Figure 2.28). The difference from line procedures arises from the fact that the packet transfer time on the network can be very long because of the transfer time on the lines and above all the waiting time in the network nodes. Under these conditions, the delay between sending a DATA packet and receipt of the corresponding acknowledgement can be very long; this severely limits the throughput on the virtual circuit. It is technically possible to overcome this limit by using a very large window. However, this approach has the disadvantage of increasing the number of packets in transit on the network and this increases the risk of network congestion. Because of this, the maximum window size W is generally limited to a very small value, which has a default value of 2 and is sometimes even reduced to 1. To limit the negative effect of the transfer time on the rate of the virtual circuit, the X.25 protocol requires the destination DTE to immediately acknowledge the DATA packets which it receives with the D bit set to 1.

When the D bit is set to 0, the DATA packets are acknowledged directly by the DCE to which the sending DTE is connected (Figure 2.29). This arrangement clearly allows a higher rate to be obtained than with end-to-end flow control, since the transfer time has only an indirect effect. In such a case, flow control can be considered as being operated by three distinct windows which correspond to DTE/DCE exchanges at the two ends and to exchanges between the two end DCEs. Flow control is thus achieved by a *back pressure mechanism* where a reduction of rate at the destination DTE/DCE interface indirectly causes a reduction of rate between the two end DCEs and then at the source DTE/DCE interface. This effect is illustrated in the example of Figure 2.29 where the network window is equal to 4 and the DTE/DCE interface windows are both equal to 2. In this figure, the DATA packets are followed in order by the send and receive numbers. RR packets are followed by the receive sequence number. It can be seen in particular that DTE B reduces the rate by slowing down acknowledgement of the DATA (2,0) and DATA (3,0) packets. As the DTE/DCE window is equal to 2, DCE B must temporarily save the DATA (4,0) and DATA (5,0) packets which have arrived. The result is a depletion of the four-packet credit granted by the network window and this in turn blocks packet transmission at the interface between DTE A and DCE A. The virtual circuit is unblocked by the acknowledgement contained in the DATA (0,4) packet which initiates immediate sending of the DATA (4,1) and DATA (5,1) packets by DCE B together with the DATA (6,1) packet by DTE A. It can be seen in this example that, with local flow control, the same packet can contain different $P(R)$ numbers at the two ends of a virtual circuit.

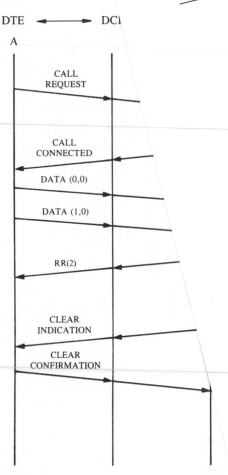

Figure 2.28 Packet exchange on an X.25 virtual circui

each packet sent, the sequence number $P(S)$ is incremen
stops when the number of packets not yet acknowledge
allowed by the window W. Transmission restarts after r
acknowledgements which are transferred by way of DAT
other station with the $P(R)$ number of the next expect
receiving DTE does not have DATA packets into which
acknowledgements, the latter must be returned by READY
packets. It can be seen that the use of sequence numbers $P($
both error checking and end-to-end flow control when the
packets is set to 1. When $D = 0$, the sequence numbers serve
control between the DTE and the network.

During the text transfer phase, the RECEIVE NOT REAL
packet can be used to acknowledge the packets already receive
that the station no longer wishes to receive new packets for some

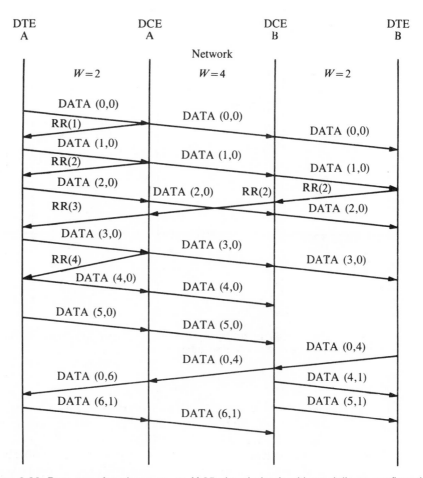

Figure 2.29 Data transfer phase on an X.25 virtual circuit without delivery confirmation

When there is no delivery confirmation, end-to-end flow control is achieved only indirectly. This is effective only if the window sizes are small. The same requirement arises here as in the case of delivery confirmation, although for different reasons.

Some networks, as an option, permit different maximum packet sizes at the two ends of the virtual circuit. In this case, the network can perform segmentation and reassembly of packets so that their size conforms to the specified limits. This type of operation is illustrated in Figure 2.30 where the maximum packet size at the DTE A/DCE A interface is larger than that of packets exchanged at the DTE B/DCE B interface. In this case, a single large packet DATA (2,1) can be delivered at the other end in the form of three successive packets DATA (2,1), DATA (3,1) and DATA (4,1) with the M bit set to 1 in the first two packets to indicate that they belong to a sequence.

When the text transfer is finished, the two correspondents must release the virtual circuit in order to free the corresponding resources. Disconnection is

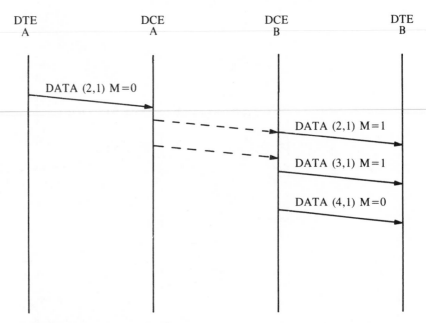

Figure 2.30 Example of segmentation when the maximum packet size is not the same at the two ends of the virtual circuit

initiated by either of the two correspondents sending a CLEAR REQUEST packet which contains an additional field specifying the reason for releasing the circuit (Figure 2.31). The CLEAR REQUEST packet is sent to the other correspondent in the form of a CLEAR INDICATION packet and the two types of packet are acknowledged by a CLEAR CONFIRMATION packet (Figure 2.28). The possible causes of releasing the virtual circuit which are reported in the CLEAR CONFIRMATION packet are indicated in Table 2.7.

In practice, clearing of the virtual circuit (level 3) is not generally followed by clearing of the data link (level 2). It has been seen that several logical channels are usually multiplexed on the link. Clearing of the logical link, indicated in Figure 2.22, only arises when the DTE discontinues all traffic with the network.

In describing establishment of the virtual circuit, it has been assumed so far that the call progressed normally throughout; this results in the return of a CALL CONNECTED packet to the caller. When the establishment attempt fails at the called DTE, for example because the latter is busy, the called subscriber produces a CLEAR REQUEST packet which causes the calling subscriber to receive a CLEAR INDICATION packet in which a field specifies the cause of clearing. This qualified reject mechanism also provides the two users with a means of negotiating the optional services used during communication as will be seen in the following section.

Notice that choice of a logical channel is performed by the DTE on the calling side and the DCE on the called side. It can thus happen that there is a *collision* when a DTE and a DCE simultaneously choose the same logical channel, one

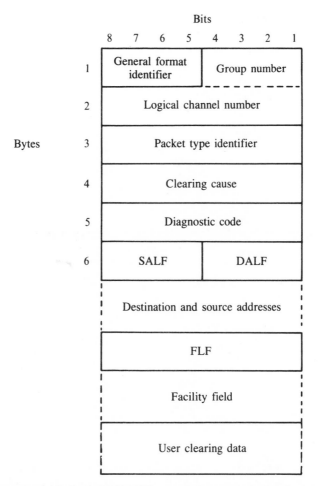

Bits

Figure 2.31 CLEAR REQUEST and CLEAR INDICATION packets

for an incoming call and the other for an outgoing call. In this case, X.25 specifies that the logical channel should be allocated to the outgoing DTE call. In practice, the chance of collision is quite small, since the protocol imposes a choice of the highest logical channel number in the DTE–DCE direction and the smallest in the DCE–DTE direction.

Subscribers can exchange urgent data outside the regular DATA packet flow by means of non-sequential INTERRUPT packets which contain up to 32 bytes of user data. These packets are transferred with high priority and are acknowledged by INTERRUPT CONFIRMATION packets. At a given time, there is never more than one INTERRUPT packet in transit on the virtual circuit.

2.6.6 Optional facilities

The X.25 protocol provides a number of *optional facilities* which can be established at the time of installation or negotiated on establishment of the virtual circuit by

Table 2.7 Significance of the diagnostics in the 'reason for clearing' field of CLEAR INDICATION packets.

Diagnostic code name	Cause of clearing
DTE originated	Call refused or virtual circuit released by distant DTE
Number busy	The called subscriber is busy
Out of order	The called subscriber is out of order
Remote procedure error	Procedure error at the interface with the called subscriber
Reverse charging acceptance not subscribed	The called subscriber refuses to pay for the call
Incompatible destination	The called system cannot operate with the service requested
Fast select acceptance not subscribed	The system cannot operate with fast select
Invalid facility request	The optional service requested is not available
Access barred	Access to the called subscriber prohibited (closed user group)
Local procedure error	Procedure error at the interface with the network
Network congestion	Network congestion
Not obtainable	No subscriber at the called number
RPOA out of order	Private equipment recognized by the administration out of order

means of a service specification field of the CALL REQUEST packet. The services to be described here must be provided by all networks which support an X.25 interface. The list of optional facilities is given in Table 2.8.

Closed user group

This service must be established by prior agreement with the network administration. It defines a group of users who can communicate with each other without limitations, but can neither call a user outside the group nor be called by a subscriber outside the group. The closed user group is thus a private network which makes use of the transport facilities of the public network. The calling DTE specifies the closed group of users with which it wishes to operate in the service specification field of the CALL REQUEST packet. The closed user group number is sent to the called DTE by the INCOMING CALL packet. Attempts to access the closed user group by an unauthorized user are blocked by the network. The latter returns a CLEAR INDICATION packet with the message 'access barred'. The closed user group service contains many variants which, for example, offer the facility of access to the open part of the network (output access) or accepting calls from the open part of the network (input access).

Flow control parameter negotiation

The selection of flow control parameters on establishment of the virtual circuit can be provided by prior agreement with the network administration. In the case where this option is used, it is possible to choose the window W and the maximum size d of the packet field separately on each logical channel. The calling subscriber indicates the values which it has chosen for the two parameters in the facility field of the CALL REQUEST packet ($d = 16, 32, 64, 128, 256, 512, 1024, 2048$ or 4096 bytes; $W = 1$–7 for modulo 8 numbering). In the case where no parameters are selected, the default values are $d = 128$ and $W = 2$. If the parameter values proposed by the calling subscriber do not suit the called subscriber, the latter can change them by inserting a counter proposal for the new values in the CALL ACCEPTED packet. In this case, the new proposals must be closer than the previous ones to the default values of $d = 128$ and $W = 2$; this enables the negotiation to complete successfully.

One-way logical channel

This type of service is established by prior agreement with the network administration and limits the virtual circuit to one-way operation.

Table 2.8 Optional facilities of X.25 networks.

On-line recording of optional facilities
Extended packet sequence numbering
D bit modification
Packet retransmission (used of the REJECT packet)
Incoming calls barred
Outgoing calls barred
One-way logical channel outgoing
One-way logical channel incoming
Non-standard default packet sizes
Non-standard default window sizes
Default throughput classes assignment
Flow control parameter negotiation
Throughput class negotiation
Closed user group (with variants)
Fast select acceptance
Reverse charging
Local charging barred
Identification of network user
Charging information
RPOA selection (selection of a private network)
Call rerouting
Notification of address change on the calling line
Notification of call rerouting
Selection and indication of transfer delay

Incoming or outgoing calls barred

This class of service, which applies to all DTE/DCE interface virtual circuits, prohibits either incoming calls or outgoing calls.

Throughput class

The virtual circuit throughput can be negotiated at the time of circuit establishment after agreement with the administration. It is clearly necessary that a request for a high throughput is consistent with requests concerning the flow control parameters and the delivery confirmation bit D.

On-line recording of user facilities

This optional service is established by prior agreement with the administration. It enables the user to request or cancel an optional facility at any time or to obtain the present state of the optional facilities which are offered. The user expresses his requirements by sending a RECORD REQUEST packet to which the DCE responds with a RECORD CONFIRMATION packet.

2.6.7 Restart and reset

Errors and impairments which affect communication across the network are treated according to the following general principles:

(a) Procedure errors which occur during establishment or clearing of the virtual circuit are reported to the DTE by clearing of the virtual circuit.
(b) Procedure errors which affect data transfer cause reset of the virtual circuit.
(c) Reset and restart packets contain a diagnostic field which provides additional information on the problem.
(d) Deadlocks are resolved by timers.
(e) The protocol must take account of procedure errors which result from incorrect interpretation by the DTE of optional facilities provided by the DCE.
(f) Actions undertaken by the DTE which depend on the received packet and the state of the DTE/DCE interface are defined by state tables.

In the case of a fault on a virtual circuit, a DTE can reset the virtual circuit by sending a RESET REQUEST packet (Figure 2.32) which contains a field giving the cause of failure. The RESET REQUEST packet must be acknowledged by a RESET CONFIRMATION packet (Figures 2.33 and 2.34). Reset causes the loss of all the packets in transit on the virtual circuit and resets the latter into the state it was in at the time of establishment; in particular, the sequence numbers are reset to zero. It can be seen immediately that the loss of sequence numbers makes it impossible for the DTE to determine the packets which have been lost during reset. Restart must therefore be provided by a higher level layer, the transport layer, in order to be sure that no packets are lost.

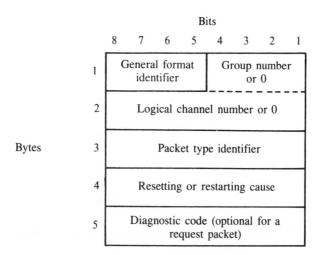

Figure 2.32 Format of RESET and RESTART REQUEST and INDICATION packets

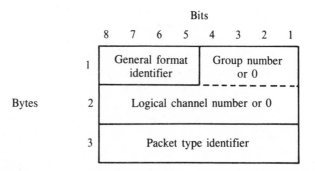

Figure 2.33 Format of RESET and RESTART CONFIRMATION packets

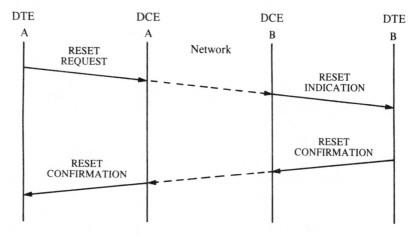

Figure 2.34 DTE reset initiative

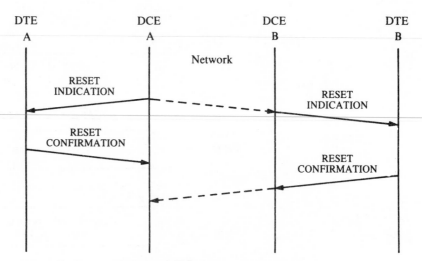

Figure 2.35 Network reset initiative

Reset can also be caused by a DCE following reset requested by the remote DTE (Figure 2.34) or following an impairment on the network (Figure 2.35). In this case, the DCE sends a RESET INDICATION packet (Figure 2.32) in which the cause field indicates the type of impairment (Table 2.9) with additional information carried by a diagnostic code field.

A *restart* procedure can be used in the case of a major fault, for example following a network node going out of service. This procedure is initiated by a DTE with a RESTART REQUEST packet (Figure 2.32) which is acknowledged with a RESTART CONFIRMATION. Restart resets all the virtual circuits of the DTE/DCE interface. It therefore has a global effect which is equivalent to clearing all the switched virtual circuits between the DTE and the DCE and

Table 2.9 Significance of the 'cause of reset' field of the RESET INDICATION packet.

Name of code	Cause of reset
DTE originated	Reset by remote DTE
Out of order	Failure of remote DTE on a permanent virtual circuit
Remote procedure error	Procedure error on the remote DTE/DCE interface
Local procedure error	Procedure error at the local DTE/DCE interface
Network congestion	Temporary network congestion
Remote DTE operational	The remote DTE is again operational (permanent virtual circuit)
Network operational	The network is again operational after temporary congestion
Incompatible destination	The interface with the remote DTE is not
Network out of order	compatible with the proposed service

resetting all the permanent virtual circuits on the same interface. Restart can also be performed on the request of the network by sending a RESTART INDICATION which must be acknowledged by the DTE with a RESTART CONFIRMATION packet (Figures 2.33 and 2.34).

For some errors, corresponding, for example, to the case of a packet carrying an unassigned logical channel number, reset or restart is too drastic. Some X.25 networks report errors of this type using DIAGNOSTIC packets whose role is simply to signal the error.

The X.25 protocol uses various timers to monitor network resources and to prevent deadlocks. Monitoring of virtual circuit establishment is performed by a timer whose guard time is set to 3 minutes and which is set by the DCE at the beginning of transmission of an INCOMING CALL packet. The called subscriber must respond before expiry of the guard time otherwise the call is cleared.

The DCEs are also provided with timers which are set to 60 seconds and are initiated when sending RESET, CLEARING and RESTART INDICATION packets. If a CONFIRMATION packet is not received before expiry of the guard time, the DCE sends a DIAGNOSTIC packet to the DTE and starts the restart or reset procedures without attempting another INDICATION packet transmission.

2.6.8 Fast select service

X.25 networks offer a datagram-like service which is known by the name '*fast select*'. This service offers the facility of exchanging up to 128 text bytes during the establishment and clearing phases of a virtual circuit and appears to the user as an optional service which he requests by specifying it in the service field of the CALL REQUEST packet.

The packet format is the same as in the case of a virtual circuit service; use is made in this case of the user data field which can amount to 128 bytes in CALL REQUEST, INCOMING CALL, CALL ACCEPTED, CALL CONNECTED, CLEAR REQUEST and CLEAR INDICATION packets. This permits up to 128 bytes of text to be transferred at the time of circuit establishment by means of the CALL REQUEST and INCOMING CALL packets without having to wait for the CALL CONNECTED response as in the case of the regular virtual circuit procedure. The called subscriber can respond to the call by returning a short message containing a maximum of 128 bytes incorporated into a CLEAR REQUEST packet. In this case, the virtual circuit is broken and the transaction reduces to the exchange of a request followed by a response, according to a procedure similar to that which corresponds to sending a datagram with a qualified acknowledgement response (Figure 2.36). In contrast, when the transaction involves more than a simple enquiry/response, the called subscriber can respond to the CALL with a CALL ACCEPTED packet which contains the first 128 bytes of the response. The rest of the communication is provided by regular text packet traffic on the virtual circuit (Figure 2.28). In this case, the fast select service serves merely to speed up the exchanges by sending piggybacked text in the call establishment packets.

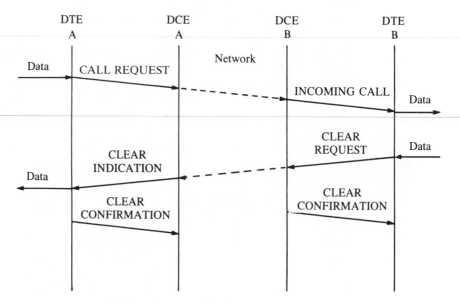

Figure 2.36 Data transfer by the fast select service

2.6.9 Packet assembly and disassembly service (PAD)

Implementation of the X.25 protocol is relatively complex, so that it is economically justified only for computers, microcomputers and intelligent terminals. Asynchronous terminals are very common and it is, therefore, necessary to provide a facility to connect them to public packet switching X.25 networks. As seen in § 2.6.1, this connection can be realized by means of auxiliary *packet assembly and disassembly* equipment (PAD) which provides protocol conversions between asynchronous terminals and the X.25 network and whose use is shared by several terminals.

Asynchronous terminals, which are called DTE-C (character mode), are connected through the network to a computer whose interface is of the X.25 type (Figure 2.20). The computer operates in packet mode and is in this case called DTE-P. The PAD serves essentially to manage the X.25 protocol on behalf of the asynchronous terminals. As the latter can be of various types, PAD operation may be defined by a number of *parameters* which allow the PAD to be compatible with the terminals. More precisely, the main functions provided by the PAD are as follows:

(a) Assembly of DTE-C characters into packets
(b) Disassembly of the user data field of packets received from the DTE-P into characters destined for the DTE-C
(c) Establishment and clearing of virtual circuits
(d) Control of reset and interrupt procedures
(e) Generation of service signals
(f) Control of packet transmission to the DTE-P

(g) Control of character transmission to the DTE-C

(h) Processing of a *break signal* from an asynchronous terminal

(i) Editing in the DTE-C

The CCITT has specified the PAD functions in Recommendation X.3, the protocol for exchanges between a DTE-C and the PAD in Recommendation X.28 and the protocol for exchanges between a DTE-P and the PAD in Recommendation X.29 [2.28].

The asynchronous terminal can be connected to the PAD by a leased line or by the switched network. In the most frequent case, where the connection uses the public telephone network, the circuit can be equipped with V.21 modems up to 300 bit/s, V.22 bis or V.23 modems for rates up to 1200 bit/s or V.23 modems with a 75 bit/s return channel for the terminal-PAD direction and transmission at 1200 bit/s for the PAD-terminal direction. The circuit must be established to conform to Recommendation V.25. If the connection is made by way of a public data network, the interface must conform to Recommendation X.20 or X.20 bis.

Commands and data are exchanged between the terminal and the PAD in asynchronous mode using characters which contain seven effective bits and one parity bit and are coded in accordance with International Alphabet No 5.

To communicate with a remote computer, the asynchronous terminal operator starts by dialling the PAD call number and setting its modem into data mode [2.36]. The operator then keys one or more periods followed by carriage return (.⟨CR⟩) in order to signal a service request to the PAD. These characters are used by the PAD to determine the bit rate of the terminal. In response to this signal, the PAD returns a prompt message which contains, for example, the network name and address of the terminal:

TELEPAC: 1 532 8981

After these exchanges, the PAD/DTE-C link is in *command mode* where all characters appearing in columns 2 to 7 of International Alphabet No 5, except SP, DEL and +, are interpreted as commands intended for the PAD and not the remote user. This mode permits a local dialogue between the asynchronous terminal and the PAD. This dialogue consists of *commands* which are sent to the PAD to which the latter responds with *indications*. The commands are terminated by the carriage return character ⟨CR⟩ which serves to delimit the end.

In command mode, operation of the PAD can be set for a particular terminal by the setting of 18 *parameters* whose role is indicated in Table 2.10. Each of these parameters is identified by a number between 1 and 18 and the particular values of each parameter are specified by a number. The set of parameter values relating to a terminal form the *terminal profile* as seen by the PAD, and the PAD keeps a distinct profile for each asynchronous input line. To permit establishment of communication, the PAD sets the parameters to the default values which constitute the *initial standard profile* at the time of calling.

In command mode, the asynchronous terminal can read or set the PAD parameters either individually or globally by selecting a *standard profile* (Table 2.11). The terminal can also control establishment and clearing of the virtual circuit or send an interrupt packet. In this way, for example, setting of parameters

Table 2.10 PAD parameters.

Parameter number	Description of parameter	Significance and values of parameter
1	Choice of escape character (change from 'data' mode to PAD command mode)	0: No escape; 1: DLE character; 32–126: Escape by corresponding character
2	Return an echo	0: No echo; 1: Echo
3	Choice of packet forwarding signal	0: No packet fowarding signal 1: Carriage return character (CR) 126,6,18: Other characters
4	Choice of packet forwarding timer delay	0–255 in 1/20 seconds
5	Terminal flow control by the PAD	0: No control; 1: Control by XON(DC1)/XOFF(DC3)
6	Service signal commands from the PAD to an asynchronous terminal	0: No service signal transmission 1: Service signal transmission 5: Confirmation message at connection time
7	Choice of PAD functions on receipt of a break signal	0: Nothing; 1: Interrupt; 2: Reset 8: Escape from the data transfer state 21: Discard output
8	Discard data destined for an asynchronous terminal	0: Normal delivery of output data 1: Discard output data (for debugging)
9	Padding after the carriage return character (CR)	0–7: Number of padding characters after CR
10	Carriage return	0: No automatic return 1–255: Number of characters before return
11	Terminal speed (accessible for read only by the DTE-P)	0–18: Rate between 110 bit/s and 64 kbit/s
12	PAD flow control by the terminal	0: Not used 1: Flow control by XON (DC1)/ XOFF(DC3)
13	Line feed (LF) inserted by PAD after carriage return (CR)	0: No line feed 1: Line inserted after CR 4–7: Other conditions
14	Padding after line feed (LF)	0: No padding 1–7: Number of padding characters after line feed
15	Editing	0: No 1: Yes
16	Character delete	0–127: Selected characters
17	Line delete	
18	Line display	

Table 2.11 PAD commands and the corresponding responses.

Command signal from the PAD	Function	Response from the PAD
STA	State request to determine if a virtual circuit is established	FREE or ENGAGED
CLR	Virtual circuit clearing request	CLR CONF (success) CLR ERR (abort)
PAR? (Parameters)	Read the current parameter values	PAR (list of the parameters with their values)
SET? (List of the parameters with their values)	Set the parameters indicated to the specified values and read the parameters	PAR (list of the parameters with their values) or INV (abort)
PROF (Identifier)	Choice of a standardized parameter profile	Acknowledgement
RESET	Reset of the virtual circuit	Acknowledgement
INT	Transmission of an interrupt packet	Acknowledgement
SET (List of parameters with their values)	Setting the parameters indicated to the specified values	Acknowledgement
Selection command signal from the PAD	Virtual circuit establishment request	Acknowledgement

2, 3 and 5 to the values 0, 1 and 0 respectively can be performed with the help of the SET command, using

Command: SET 2:0, 3:1, 5:0 ⟨CR⟩
Response: ⟨CR⟩⟨LF⟩

In a similar manner, the SET? command causes the parameters to be set and then read, for example:

Command: SET? 2:0, 3:1, 5:0 ⟨CR⟩
Response: TELEPAC: PAR 002:000, 003:001, 005:000

where PAR is a response to a read command.

Selection of a standard profile is made very simply, by a PROF command followed by the number of the selected profile:

Command: PROF 3 ⟨CR⟩
Response: ⟨CR⟩⟨LF⟩

After possible setting of the parameters, the operator establishes the virtual circuit by sending the network address of the DTE-P, which is the host computer, using for example

Command: 5 642 3565 ⟨CR⟩

This command causes the PAD to send a CALL REQUEST packet and, if successful, the return of a CALL CONNECTED packet; as a result of this the PAD sends an acknowledgement to the terminal from the host. In the establishment command, the called station address can be preceded by a *reverse charging* command (REV), and/or an indication of call *priority* (N – normal, P – priority). The called number may be followed by the *user data* for insertion into the corresponding field of the CALL REQUEST packet. This leads, for example, to a command of the form

    Command: REV P 5 642 3565, CMS ⟨CR⟩

After the return of the PAD response, the virtual circuit is established and the PAD/DTE-C link changes to *data mode*. The characters sent by the asynchronous terminal are then assembled into data packets which are sent to the DTE-P. Conversely, data packets received by the PAD are disassembled into characters which are sent to the terminal.

The rules used by the PAD for sending data packets can be modified by parameters 3 and 4 (Table 2.10). By default, the parameters are set so that a data packet is sent each time the number of characters received by the PAD reaches the maximum size of the data field. By selection of parameter 4, the operator can change this policy by causing a data packet to be sent whenever no character has been sent during a time which exceeds the specified guard time. Finally, the operator can decide, using option 3, to control the sending of packets himself by means of a particular character which can, for example, be ⟨CR⟩. It can be shown that this last option is the most efficient in terms of network data transfer time [2.37].

As an option with parameter 1, the operator can change temporarily from data mode to command mode by sending an escape character which is generally the ⟨DLE⟩ character. This permits the PAD parameters to be read or modified in the course of a transaction. Return to data mode is then caused by sending the ⟨CR⟩ character. This technique can be used to clear the virtual circuit with the help of the CLR (*clear*) command after changing to command mode, using for example

    Command: ⟨DLE⟩ CLR ⟨CR⟩
    Response: TELEPAC: CALL CLEARED. LOCAL DIRECTIVE

When the escape character option is not used, the virtual circuit is cleared by disconnection of the PAD access line.

Exchanges between the PAD and the DTE-P are controlled by the X.25 packet protocol. However, the DTE-P has the facility of passing commands destined for the PAD by way of packets whose Q bit is set to 1. These commands are specified by CCITT Recommendation X.29 [2.29] and they permit the PAD parameters to be read and set (Table 2.12). They are also used to clear the virtual circuit with the terminal and to transmit error messages. The service messages exchanged between the PAD and a DTE-P are contained in the user data field of X.25 packets whose Q bit is set to 1.

Table 2.12 X.25 service messages exchanged between the PAD and a DTE-P.

Message type	Direction	Function
Set	DTE-PAD	Request setting of parameters to the requested values
Set and read	DTE-PAD	Request setting and reading of the indicated parameters
Read	DTE-PAD	Request reading of the indicated parameters
Parameter indication	PAD-DTE	Indication of the value of the requested parameters
Invitation to clear	DTE-PAD	Clearing request of the virtual circuit sent to the PAD
Break indication	DTE/PAD	Break indication sent by the PAD or DTE-P
Error message	DTE/PAD	Error message sent by the DTE or the PAD to indicate an error in the previous message

2.7 OSI CONNECTION-MODE NETWORK LAYER

2.7.1 Introduction

In the context of the OSI reference model for open systems, the ISO has recently defined a connection-mode network layer [2.38]–[2.39]. In practice, the specification relates essentially to the network service since most public packet switching networks operate according to the X.25 standard previously defined by the CCITT. This has posed a compatibility problem which has been resolved only imperfectly with the 1980 X.25 protocol [2.28] and a satisfactory solution has been found only with the new version of the X.25 protocol standardized in 1984 [2.29]. The OSI connection-mode network service will be described here and it will subsequently be shown how this service can be provided with X.25 networks.

2.7.2 Connection-mode network service

The *connection-mode network service (CONS)* is provided by means of primitives which are shown with their parameters in Table 2.13. The main services provided by the network are as follows:

(a) Independence from the underlying transmission media
(b) End-to-end transfer
(c) Transparent information transfer
(d) Quality of service selection
(e) Network service user addressing

Table 2.13 Network service primitives.

Primitives	Parameters
N-CONNECT request	called address
	calling address
	receipt confirmation selection
	expedited data selection
	quality of service parameter set
	user data
N-CONNECT indication	"
N-CONNECT response	responding address
	receipt confirmation selection
	expedited data selection
	quality of service parameter set
	user data
N-CONNECT confirm	"
N-DATA request	user data
	confirmation request
N-DATA indication	"
N-DATA ACKNOWLEDGE request	
N-DATA ACKNOWLEDGE indication	
N-EXPEDITED DATA request	user data
N-EXPEDITED DATA indication	user data
N-RESET request	reason
N-RESET indication	originator
	reason
N-RESET response	
N-RESET confirm	
N-DISCONNECT request	reason
	user data
	responding address
N-DISCONNECT indication	originator
	reason
	user data
	responding address

The basic service permits *establishment of the network connection, normal data transfer, connection release* and *resetting*. Two options are provided to ensure *expedited data transfer* and *receipt confirmation*. The latter service enables a confirmation of receipt to be obtained from the remote user.

The network connection is established on the initiative of any one user by an N-CONNECT request primitive which contains the source and destination addresses as parameters. This primitive also permits receipt confirmation and expedited data options to be selected by setting the corresponding parameters. The N-CONNECT request primitive can also pass up to 128 bytes of user data as a parameter; these are transmitted to the remote user at the time of establishment of the virtual circuit. Finally, the N-CONNECT request primitive contains various *quality of service (QOS)* parameters which permit the required performance and general characteristics of the service to be specified.

The various quality of service parameters are shown in Table 2.14. The *network connection establishment delay* parameter defines the maximum delay between the sending of an N-CONNECT request and receipt of an N-CONNECT confirm primitive which indicates that the virtual circuit has been successfully established. The *network connection establishment failure probability* parameter specifies the maximum acceptable ratio of the total number of connection establishment failures to the total number of establishment attempts.

The quality of service of the network during the normal data transfer phase can be defined by the following parameters: *throughput, transfer delay, residual error rate, transfer failure probability* and *probability of a network connection break* (network connection resilience). It will be seen in the following chapter that these parameters play a very important role when the transport layer selects the protocol class which ensures reliable end-to-end service with the required data rates and response times. The *residual error rate* is the ratio of the number of incorrect, lost or duplicated packets to the total number of packets transferred across the service boundary of the network. It consists in this case of *'errors' not detected by the network*. The *probability of a network connection break* characterizes the frequency of errors signalled by the network and originating from it. These errors correspond either to a connection release which is announced by an N-DISCONNECT indication primitive or a connection reset which is indicated by an N-RESET indication primitive. It will be seen in the following chapter that the transport protocols are very different when the network has, and does not have, an acceptable residual error rate and a low probability of a network connection break.

During the establishment phase of the network connection, the final selection of options and parameters results from a tripartite negotiation between the two users and the provider of the network service. The values indicated by the

Table 2.14 Quality of service parameters.

Phase	Quality of service parameters
Network connection establishment	Network connection establishment delay
	Establishment failure probability
Data transfer	Throughput
	Transfer delay
	Residual error rate
	Network connection resilience
	Transfer failure probability
Network connection release	Network connection release delay
	Network connection release failure probability
Performance of the network service	Network connection protection
	Network connection priority
	Maximum acceptable cost

parameters of the N-CONNECT request primitive represent only an initial proposal which is liable to be modified by the remote user or the provider of the network service. The possible counter proposals to the initial requests of the caller are chosen in accordance with pre-arranged rules; they will usually have to be closer to the default values than the proposals. They are contained in the N-CONNECT indication, response and confirm parameters (Figure 2.37).

The *performance* of the network service can be specified by three parameters. The *network connection protection* parameter characterizes the option of security protection with the following possibilities:

(a) No protection
(b) Protection against passive monitoring
(c) Protection against malicious attempts at modification, replay, addition or deletion
(d) Protection against both types of attack

The *network connection priority* parameter specifies the relative importance of the various network connections, that is the order in which network connections can be broken or subjected to a degradation of service quality. This priority does not concern the order in which the data packets are transmitted by the network since the problem of routing urgent packets is taken care of by the expedited data service. It is also possible to specify the *maximum acceptable cost* of a network connection by means of a parameter which can be expressed in units of absolute or relative cost.

Normal data transfer is performed using N-DATA request and indication primitives which contain the block of user data as a parameter. This block can be of any size provided that it is a multiple of a byte. It represents the *network service data unit (NSDU)* which is exchanged at the interface between the network and transport layers. When the length of the data block exceeds the maximum

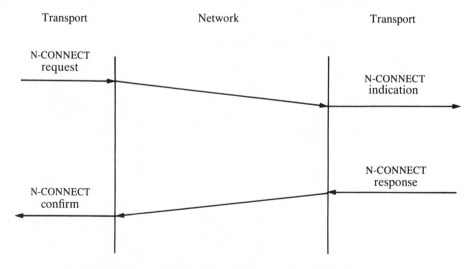

Figure 2.37 Establishment phase of network connection

size allowed by the network, the NSDU is segmented to form several *network protocol data units (NPDU)* which, in this case, are the network data packets (Figure 2.38). The N-DATA primitives also include a *delivery confirmation* parameter which enables the sender to check that the data has indeed been delivered to the called address. When confirmation is requested, the remote user must acknowledge receipt of the received NSDU by means of a parameterless N-DATA ACKNOWLEDGE request primitive which in turn causes an N-DATA ACKNOWLEDGE indication primitive to be notified to the sender.

Urgent data can be exchanged at the network service interface by N-EXPEDITED DATA request and indication primitives if the corresponding option was selected at the time of establishment. These primitives contain only one user data parameter and the length of the data block must not exceed 32 bytes.

The N-DISCONNECT request and indication primitives serve to clear the virtual circuit and can include the reason for clearing as a parameter; the indication primitive can contain the *originator*, which can be the network itself, the remote user or 'unknown'.

The N-RESET request, indication, response and confirm primitives provide a reset service which is used by the network to indicate a loss of packets not involving breaking of the connection (an indicated incident). The reset service is also used by a network service user to resynchronize the network connection.

2.7.3 Compatibility with X.25 networks

As most public packet switching networks are of the X.25 type, it is necessary that the OSI network service should be compatible with the X.25 packet protocol.

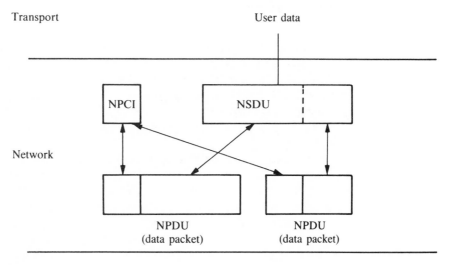

Figure 2.38 Segmentation and reassembly in the network layer level

The 1980 version of this protocol [2.28] does not permit complete compatibility, particularly in respect of quality of service and expedited data transfer parameters. In order to resolve this problem and to extend the facilities of X.25 networks, the CCITT has made significant modifications to the initial protocol. This was formalized by the publication of protocol X.25 1984 [2.29]. The new version of the protocol contains an extended range of optional facilities and it permits an expedited data service to be provided by INTERRUPT packets due to extension of the data field of these packets from 1 to 32 bytes maximum.

The main modifications of the 1980 version by the 1984 version of the X.25 protocol at packet level are as follows [2.32]:

(a) Elimination of the datagram service
(b) The user field of INTERRUPT packets is extended to a maximum of 32 bytes (expedited data service)
(c) Mandatory fast select service with a CALL REQUEST packet data field extended to 128 bytes
(d) Maximum size of the facility field of CALL REQUEST, INCOMING CALL, CALL ACCEPTED and CALL CONNECTED packets extended to 109 bytes
(e) Introduction of a supplementary transfer time selection service (quality of service)
(f) Extension of closed user group facilities
(g) Optional negotiation of throughput, transfer time, address extension and expedited data service by the DTEs
(h) Introduction of a *recording* procedure which permits reading and on-line modification of some optional facilities which are normally provided by subscription
(i) Optional selection of the transit network on which the circuit must be established
(j) Introduction of new facilities such as *group lines* and call transfer
(k) Improvement of error processing procedures

Use of an X.25 network with the OSI network service is specified by a provisional ISO standard published in 1986 [2.39]. This standard provides for connection to 1984 X.25 networks and also the possibility of use with 1980 X.25 networks with a more limited service. In the latter case, it is necessary to add a matching *subnetwork dependent convergence protocol (SNDCP)* to the X.25 packet protocol. Use of the ISO service with 1984 X.25 networks will be described here briefly.

The mapping between the service primitive parameters and the fields or facilities of X.25 packets is indicated in Table 2.15. Establishment of the service starts with an N-CONNECT request primitive which causes a CALL REQUEST packet to be sent; its facility field is coded in accordance with the parameters contained by the primitive (Figure 2.39). As the call data contained by the primitive can have a length up to 128 bytes, the network must operate the fast select service. The user data field of normal CALL REQUEST packets is limited to 16 bytes but it can be 128 bytes with the optional fast select service. Similarly, the arrival of an INCOMING CALL packet is notified by an N-CONNECT indication

Table 2.15 Mapping between network service primitive parameters and 1984 X.25 supplementary packet services.

Parameters of network service primitives	Fields and facilities of 1984 X.25 packets
Called address	Destination address field
	Address extension field
Calling address	Source address field
	Address extension field
Responding address	Destination address field
	Address extension field
Receipt confirmation selection	General format identifier (bit D)
Expedited data selection	Supplementary negotiation service for expedited data
Quality of service parameters	Supplementary negotiation service for throughput class
	Supplementary service for end-to-end transfer delay
	Supplementary service for minimum throughput class
	Supplementary service for transfer delay choice
	Supplementary negotiation service for end-to-end transfer delay
User data	User data field
	Supplementary fast select service

primitive, and the N-CONNECT response and confirm primitives correspond to CALL ACCEPTED and CALL CONNECTED packets respectively.

Similarly, the N-DISCONNECT request and indication primitives correspond to CLEAR REQUEST and INDICATION packets respectively. In contrast, the CLEAR CONFIRMATION packets are not associated with the network service primitives and their significance here is purely internal to the network (Figure 2.40).

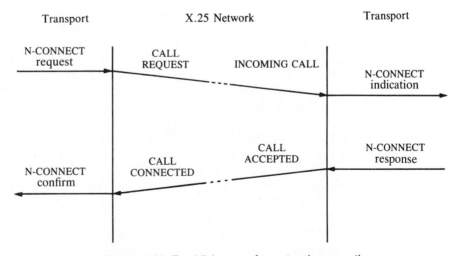

Figure 2.39 Establishment of a network connection

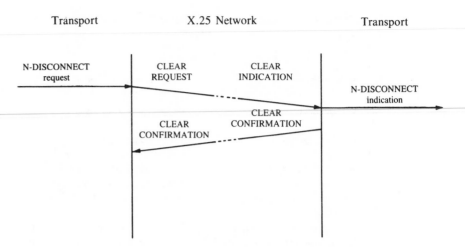

Figure 2.40 Clearing of a network connection

When the receipt confirmation option is not selected, sending an N-DATA request by the user causes a DATA packet to be sent; this contains the data passed as a parameter of the primitive in its user data field. The D bit of the DATA packet must be set to 0 to indicate that no receipt confirmation is requested from the remote user. As the length of the data block passed by the primitive is arbitrary, while the size of the DATA packet user data field is limited, the network can make several successive DATA packets correspond to a single N-DATA request primitive. In this case, all packets have their M bit at 1 except the last of the series whose M bit is at 0. At the receiver, the receipt of the last DATA packet with the M bit at 0 causes transmission of an N-DATA indication primitive which contains the complete block of user data as a parameter.

When the N-DATA request primitive indicates that the receipt confirmation option has been chosen, the D bit of the last DATA packet of each group is set to 1 to call for a receipt confirmation by the correspondent (Figure 2.41). The remote user must respond immediately to an N-DATA indication primitive with an acknowledgement which takes the form of a parameterless N-DATA ACKNOWLEDGE request primitive. This last primitive is not associated with a particular X.25 packet since the acknowledgement is actually provided from the receive number $P(R)$. The remote network entity must respond to the primitive by sending a DATA packet or, by default, an RR or RNR packet whose sequence number $P(R)$ acknowledges the received data. At the transmitter, receipt of the packet causes notification of a parameterless N-DATA ACKNOWLEDGE indication primitive which terminates the exchanges.

Urgent data is transferred by means of an N-EXPEDITED DATA request primitive whose user data parameter can contain up to 32 bytes. This primitive causes an INTERRUPT packet to be sent whose arrival at the other end of the link leads to sending of an N-EXPEDITED DATA indication primitive (Figure 2.42). There is no segmentation here and no receipt confirmation option.

Reset can be performed on the initiative of one of the users, by an N-RESET request primitive which results in the exchanges indicated in Figure 2.43 with a

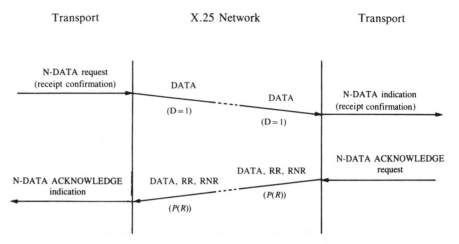

Figure 2.41 Data transfer with receipt confirmation

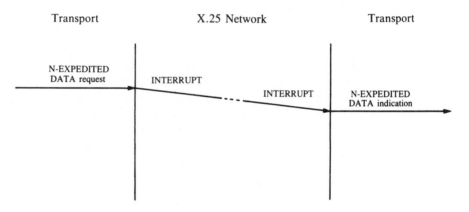

Figure 2.42 Expedited data transfer

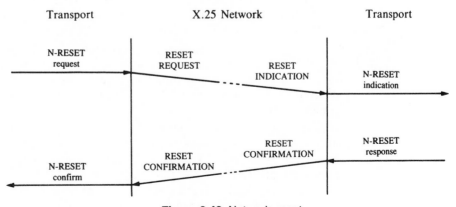

Figure 2.43 Network reset

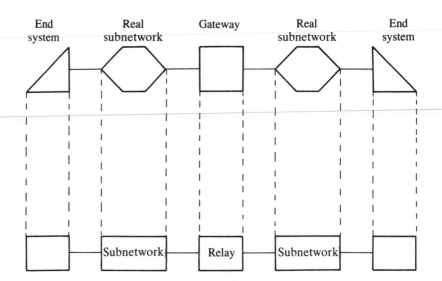

Figure 2.44 Example of the connection of two systems across two real subnetworks

direct mapping between the RESET primitives and the corresponding packets. It should be noted, however, that the N-RESET confirm primitive is used here only when the transport entity wishes to indicate to the user that it is ready to resume data transmission after reset. The network can take the initiative for a reset or recovery which leads to the arrival of a RESET INDICATION or RESTART INDICATION packet respectively at the ends of the link. These events are signalled to the user by an N-RESET indication primitive containing the origin and the cause as parameters.

2.7.4 Internal organization of the network layer

It was shown in the previous section that it is necessary to add a supplementary protocol layer to the 1980 X.25 packet protocol in order to provide, at least partially, an OSI network service. In practice, the problem of adapting the network layers arises in a much more general manner particularly in the case where the end systems are connected by way of one or more real subnetworks. This type of situation is illustrated in Figure 2.44 where two end systems are connected by way of two real subnetworks of which one could, for example, be a public packet switching network and the other a local area network. The two subnetworks are in this case interconnected by an intermediate *relay system* which could for example be an *interworking unit* or *gateway*. The upper part of Figure 2.44 represents the real components of the complete system and the lower part indicates the abstract objects to which they correspond.

With a system such as that of Figure 2.44, difficult compatibility problems can arise between the network protocols operated by the various subgroups. To solve this type of problem, the ISO has recently proposed [2.40] a model of the internal organization of the network layer with the following objectives:

(a) To simplify the use of network protocols to provide a network service in different situations
(b) To limit the proliferation of network protocols with redundant functions
(c) To clarify the needs and provide a guide for the development of network protocol standards

The structure proposed by the ISO for the network layer is composed of three sublayers (Figure 2.45). At the lowest level, the *subnetwork access protocol (SNACP)* provides the network functions which are specific to a particular real subnetwork and which particularly concern routing in the subnetwork. As the service provided by this sublayer is not generally that which is required, it is most often necessary to match the SNACP protocol by means of two additional sublayers. At the highest level, this adaptation function is taken care of by a *subnetwork independent convergence protocol (SNICP)* which provides an OSI network service at the transport layer and makes use of a well-defined set of functions provided by an intermediate protocol called the *subnetwork dependent convergence protocol (SNDCP)*. It can be seen that the SNICP protocol is relatively independent of the characteristics of the real subnetwork and the SNDCP protocol is essentially to provide decoupling between the characteristics which are associated with the subnetwork and the organization of the protocol which provides the network service to the transport layer.

In practice, the relative importance of the three sublayers depends very much on the real subnetwork which is used. In particular, if the SNACP protocol provides an ISO service directly, the SNDCP and SNICP protocols are not required.

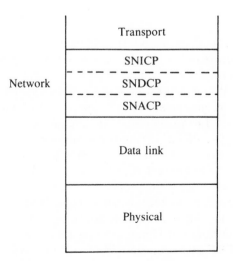

Figure 2.45 Internal organization of the network layer

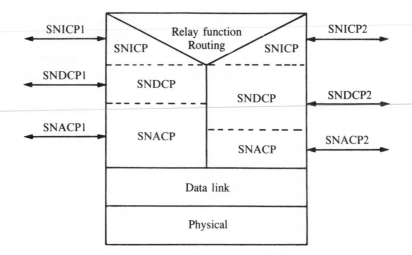

Figure 2.46 Relay at the network level

When the interconnection between two real subnetworks is realized in the network layer, the gateway can be organized as shown in Figure 2.46 where the SNDCP and SNICP sublayers which are located on each side match the protocols to the two subnetworks with a common ISO service. The function of the relay which is located in the upper part of the gateway is to permit information received from a user in one of the networks to be sent to a user in the other network. The gateway also provides routing between the networks.

2.8 CONNECTIONLESS OSI NETWORK LAYER

2.8.1 Overview

Network protocols with connection are almost universally used with long-haul networks but they are badly suited to the environment of a high-speed local area network where the short response times lead to the use of connectionless services with connection establishment possibly relegated to the transport layer. In response to requirements of this type, the ISO has established provisional standard ISO 8473 for a *connectionless mode network service (CLNS)* [2.41]–[2.44].

The ISO 8473 standard defines only the network sublayer which is independent of the real subnetwork. Hence it consists of an SNICP protocol, as presented in the previous section, which forms only the upper part of the network layer (Figure 2.45). This protocol must, in most cases, also contain an SNACP access sublayer particular to the real subnetwork and a matching SNDCP sublayer.

The general organization of protocols with a connectionless network layer is illustrated in Figure 2.47 for the case of a local area network operated with the IEEE 802 protocols. The data link layer is here divided into a *medium access control (MAC)* sublayer which is specific to the network type and a *logical link control (LLC)* sublayer which includes the functions which are independent of

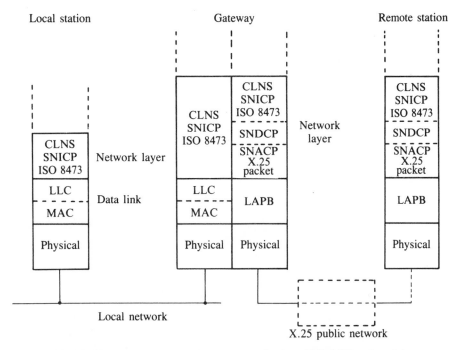

Figure 2.47 Example of use of the connectionless ISO 8473 network layer

network type. In its simplest version, called Type 1, the LLC data link protocol is connectionless with data transfer performed by isolated frames without sequence numbering. The service provided in this way is almost that which is used by the ISO 8473 SNICP protocol apart from some details concerning quality of service parameters. This allows all the network layer functions to be provided with the single ISO 8473 protocol.

In the example of Figure 2.47, the local network stations are able to access a public X.25 network by way of a gateway. The SNACP network sublayer is in this case the X.25 protocol at packet level. As the latter, in its 1984 version, no longer provides a datagram service, it is necessary to use an SNDCP matching protocol which provides a connectionless service to protocol ISO 8473 from the X.25 service with connection. This configuration is clearly rather awkward but is justified in this case by the need to use existing public networks.

2.8.2 Connectionless network service

The connectionless network service reduces essentially to the transfer of independent network service data units (NSDU). The NSDU is passed to the network as a parameter of an N-UNITDATA request primitive whose other parameters specify the required quality of service and the calling and called addresses (Table 2.16). The NSDU is delivered to the network layer user by an N-UNITDATA indication primitive which contains the same parameters but with

Table 2.16 Connectionless network. Service primitives.

Primitives	Parameters
N-UNITDATA request	Calling address
	Called address
	Quality of service
	User data
N-UNITDATA indication	”
N-FACILITY request	Service characteristics
	Quality of service parameters
N-FACILITY indication	Called address
	Service characteristics
	Quality of service parameters
	Report reason
N-REPORT indication	”

values which can be different for the quality of service (Figure 2.48).

In this case there cannot be negotiation of service quality between the users since there is no communication establishment phase. This means that the users can only select the quality of service parameters, taking account of a prior agreement with the network service provider. In this context, it is necessary to offer information on the values of service quality parameters offered by the network to users of the network layer. To this end, the user can send an N-FACILITY request primitive which contains the list of quality of service parameters whose value is requested (Table 2.16). In response the network sends an N-FACILITY indication primitive which contains the values of the quality of service parameters. The network can also take the initiative of sending an N-REPORT indication primitive to indicate a quality of service degradation. This primitive contains the new values of quality of service as a parameter together with the reason for sending the report which could, for example, be network congestion.

The quality of service parameters are similar to those which are used in the connection mode service (Table 2.17). The *priority* parameter allows the network

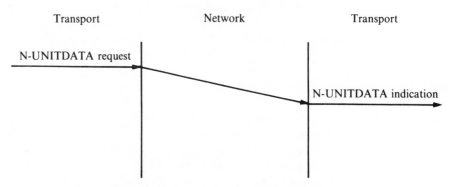

Figure 2.48 Connectionless data transfer

Table 2.17 Connectionless network. Quality of service parameters.

Transmit delay
Unauthorised access protection
Maximum acceptable cost specification
Minimum cost service selection
Residual error rate
Priority
Choice of route

to determine the order in which it must route the various NSDU. The *choice of route* parameter allows the user to select a particular route for the NSDU which is submitted to the network. In this case, the parameter contains the list of intermediate nodes through which the packet must pass. This option is useful when the user has the choice of several real subnetworks whose costs are different.

The choice of transport protocol to be used requires knowledge of certain *general network service characteristics* whose values can be given to the transport entity by the N-FACILITY indication and N-REPORT indication primitives. The *congestion control characteristic* indicates if the user is ready to accept indications from the network which signal a refusal to transmit new NSDUs following temporary congestion. If congestion occurs when this flow control option is in operation, the network is not permitted to discard the NSDU from the user. In contrast, NSDUs transmitted by users who have not selected the congestion control option can be rejected when the network is overloaded.

More generally, the connectionless network provides a datagram service with 'best-effort' transfer of isolated packets without guarantee against loss of packets, transmission errors or out-of-sequence delivery. An indication of the last type of failure can be provided by the network in the form of a *sequence preservation characteristic* which is the ratio of the number of transfers in sequence to the total number of transfers.

The last general characteristic concerns the *maximum packet lifetime*. The network assigns a maximum lifetime to packets and destroys them when their lifetime is exhausted. Knowledge of the maximum packet lifetime is required in the transport layer to determine the guard time of certain timers.

2.8.3 Connectionless network protocol

The ISO 8473 connectionless network protocol [2.41] provides a datagram service with only two types of packet, *protocol data units (PDU)*, which are *data packets (DT)* and *error packets (ER)*. It provides the service in the transport layer described in the previous section with the primitives indicated in Table 2.16. It has been seen that this protocol is of the SNICP type, that is it is independent of the type of real subnetwork used. The ISO 8473 protocol normally uses the services of a matching SNDCP protocol (Figure 2.45) which is provided with the SN-UNITDATA request and SN-UNITDATA indication primitives (Table 2.18).

Table 2.18 Service primitives of the SNDCP network sublayer.

Primitives	Parameters
SN-UNITDATA request	called address
	calling address
	quality of service
	user data
SN-UNITDATA indication	,,

These two primitives define a connectionless service with calling/called address, quality of service and user data parameters. The quality of service parameters are in principle the same as for the N-UNITDATA network service primitives but there is provision for the protocol to operate in a fallback mode if the service quality requested cannot be provided by the SNDCP sublayer and the real subnetwork. The user data parameter of the SN-UNITDATA primitives contains a *subnetwork service data unit (SNSDU)* whose length is limited by the maximum size of the data fields of the real subnetwork packets. For the system to be useful, it is clearly necessary that this size should be sufficient to contain an ISO 8473 protocol packet containing at least one byte of user data. This imposes a minimum SNSDU size of 256 bytes.

The ISO 8473 protocol entity must be able to know the time locally. This timing service is provided by sending an S-TIMER request primitive to the local environment, with parameters which specify the required guard time together with the timer identifier and possibly a subsidiary identifier if several timers with the same name are used (Table 2.19). The S-TIMER response primitive allows the local environment to indicate that the guard time of a timer has expired and the S-TIMER cancel primitive is used by the ISO 8473 protocol entity to stop one or more timers.

The general format of ISO 8473 protocol packets is shown in Figure 2.49. The packets contain user data and a *header* which contains a *fixed part* of 9 bytes together with a *variable part* formed by the source/destination address, segmentation and option fields.

Table 2.19 Timer primitives.

Primitives	Parameters
S-TIMER request	Guard time
	Timer identifier
	Timer sub-identifier
S-TIMER response	Guard time
	Timer sub-identifier
S-TIMER cancel	,,

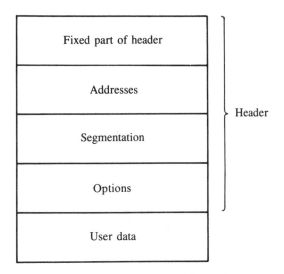

Figure 2.49 Structure of connectionless ISO 8473 protocol packets

The structure of data packets DT is shown in Figure 2.50. The fixed part of the header starts with a network protocol identifier byte whose value in this case is 1000 0001 to indicate ISO protocol 8473. This data is completed by byte 3 which identifies the protocol version and takes the binary value 0000 0001 for the standard version of ISO protocol 8473. Byte 2 indicates the header length in bytes, that is the length of the packet with the exception of the data field. This length must not exceed 254 bytes.

The lifetime byte is loaded with a particular value by the originator and decremented each time the packet passes through a network node. Every packet whose lifetime is zero must be discarded; this ensures that packets cannot jam the network by circulating indefinitely. Discarding a data packet can cause an error packet ER to be sent to the sender which signals the event to it.

Byte 5 contains a type field which distinguishes data packets DT from error packets ER. It also contains an E/R flag which can be set to 1 in data packets to request the sending of an error packet if the data packet is discarded.

It has been seen that the underlying service of ISO protocol 8473 accepts only data units of limited length. As these data units are ISO 8473 protocol packets, it can be necessary to segment messages into several DT data packets. In order to control segmentation, byte 5 contains an SP flag (*segmentation permitted*) which, when set to 1, indicates that segmentation is permitted. When segmentation is used, the MS flag (*more segments*) is set to 1 when the last byte of the user data field is not the last byte of the segment. In order to permit reassembly of packets into messages at the destination, bytes 6 and 7 contain the length of the segment which is the length of the packet in this case. The segmentation part of the header, which is used only if the SP bit is set to 1, contains identification of the message to which the packet belongs, an offset which indicates the position of the segment in the message together with the total length of the packet which will be reassembled from the set of segments. This exhaustive information permits

Bytes

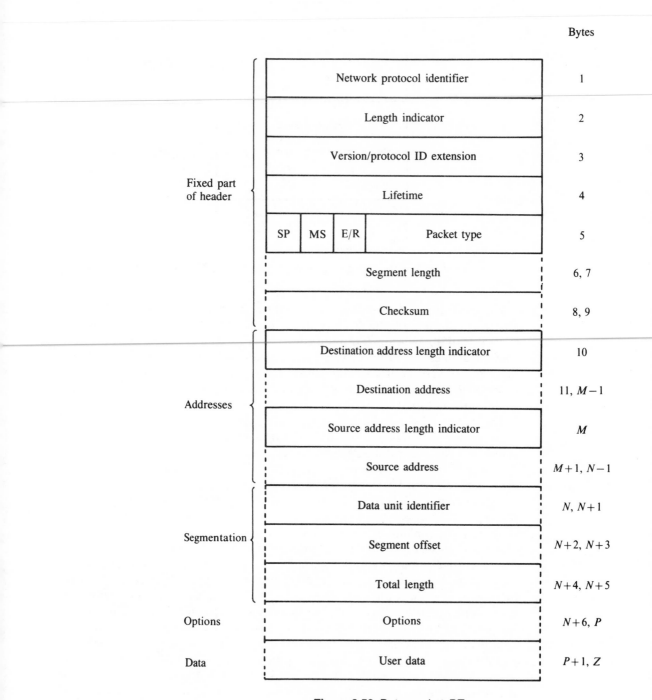

Figure 2.50 Data packet DT

segmentation and reassembly of packets at intermediate nodes when the route passes through several real subnetworks having different maximum packet sizes. The fixed part of the header finishes with two error check bytes by computing a *checksum* according to principles which will be given in the following chapter.

The variable part of the header contains the source and destination address fields, which are required since communication does not involve a virtual circuit.

The option fields of the variable part of the header contain an indication of the required quality of service together with a choice of padding, security, source routing, route recording and priority options. *Padding* offers the facility of increasing the length of packet headers with padding bits to give them a convenient size. The *security* option is used in conjunction with protective devices to deter unauthorized access. The *source routing* option is used to specify, partially or totally, the path which the packets must follow. To complement this, the *route recording* option causes recording by the packet of the addresses of the various nodes through which it passes. Finally, a last option indicates the *relative priority* of the packet.

The DT data packet finishes with a user data field whose size is the difference between the packet length, indicated by bytes 6 and 7, and the length of the header indicated by byte 2.

The *error packet (ER)* is very similar to the data packets DT (Figure 2.51). It is sent to signal the discarding of a data packet. Its destination address is thus the source address of the data packet which is discarded. Its source address indicates the address of the node which originated the discarding. The error packet contains a field which specifies the *reason for discard* which could be an incorrect checksum for example, or network congestion. To permit a more detailed analysis of the reason for discard, the error packet ER includes an *error report data field* which contains all or part of the discarded data packet.

The general organization of the connectionless ISO 8473 network layer is shown in Figure 2.52 where the functions which have already been partially presented in relation to packets are found again. The data provided by the transport layer is assembled into packets by the *PDU composition function (NIPDU)* with source routing and padding functions as options. Conversely, packets from the subnetwork are possibly subjected to reassembly before being inserted into the input queue. Their header is then checked by means of a checksum.

The packets received or assembled locally are then subjected to a *header format analysis* which serves firstly to determine which version of the protocol or which subset of the protocol must be used. At this level, analysis of the destination address permits a decision as to whether the station is the packet destination or the packet must be sent on the network; this is the case when the station is the source of the packet or when it is an intermediate node on the packet route.

The packets are then processed by an *unauthorized access protection function*, then, for those which must be transmitted or retransmitted, by a *PDU lifetime control function*. The latter function decrements the value of the lifetime and destroys packets whose lifetime is exhausted.

Packets destined for the station are then opened by the *PDU decomposition function (NIPDU)* in such a way as to provide user data in the transport layer as a parameter of an N-UNITDATA indication primitive. Retransmitted packets

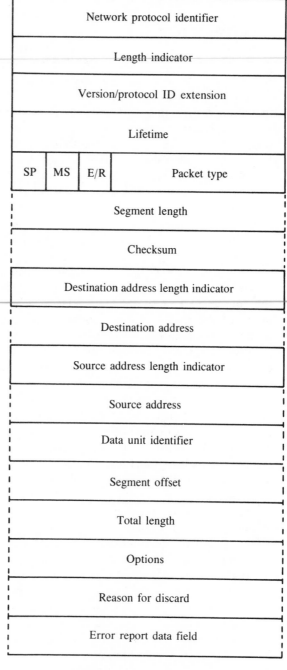

Figure 2.51 Error packet ER

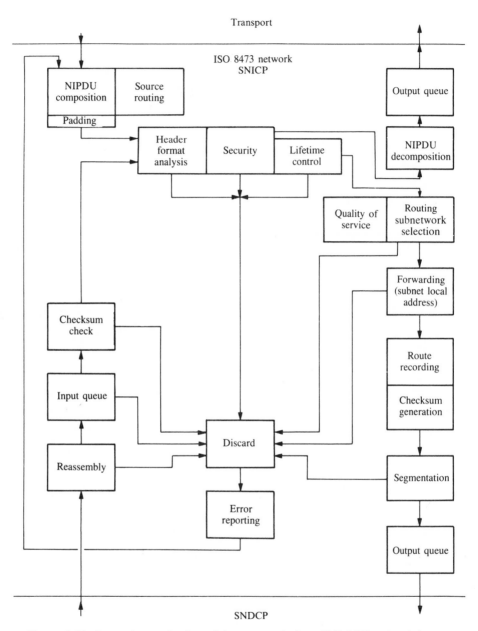

Figure 2.52 General organization of the connectionless ISO 8473 network layer

are first processed by a *route PDU function* which determines the real subnetwork on which the packet must be sent. The packets may then be segmented and the checksum for their header is computed. They are finally passed to the subnetwork or the SNDCP layer as parameters of an SN-UNITDATA request primitive.

2.8.4 Use of the ISO 8473 connectionless protocol with X.25 networks

X.25 networks offer a virtual circuit service, while the ISO 8473 protocol uses a connectionless service. The ISO 8473 protocol, therefore, can only be used with an X.25 network if a matching SNDCP sublayer is used to convert the connection mode service into a connectionless service [2.43].

The mechanism operated by the SNDCP protocol consists of making virtual circuits available to packets which are submitted to it by the ISO 8473 layer. The rule is to establish a new virtual circuit each time a new packet arrives for which no circuit is available, or each time the size of the queues for the existing virtual circuits exceeds a given limit. A new virtual circuit can also be established on the initiative of the network management service. The SNDCP protocol takes charge of the virtual circuit establishment operations in a manner transparent to the ISO 8473 network layer.

Clearing of virtual circuits is also performed by the SNDCP protocol without explicit intervention by the ISO 8473 network layer, for example on the expiry of a guard time after sending the last packet.

The SNDCP protocol must also operate a mechanism for resolving collisions which can occur when two distant SNDCP entities attempt simultaneously to establish a virtual circuit between each other.

CHAPTER 3

TRANSPORT LAYER

The ***transport layer*** is the highest level of the *low-level services*. It provides an ***end-to-end transport service*** of data between users *in a reliable and effective manner*. It lies above the physical, data link and network layers which control communication between adjacent equipment, and its principal role is to establish a relationship between the processes which reside in the user terminal equipment. At this level, the exchanges are controlled by a *transport protocol* which is sometimes called the *host-to-host protocol*. The transport layer provides a transport service for the higher layers which frees them from problems concerning routing of data and makes their operation independent of the actual means of communication used. The transport layer is implemented in the form of software which is often called a ***transport station*** and sometimes a *network control program* or *transmission control program*.

The design of the transport layer poses very difficult problems which are due essentially to the fact that the packets are routed by a network with transfer times which can vary widely. On the other hand, it is also necessary to provide mechanisms which ensure data integrity in the case of breakdown of the network or one of the hosts which participate in the communication. In this chapter, the main features of the transport layer will be presented. Subsequently the ISO transport protocols with and without connection will be described [3.1–3.21].

3.1 GENERAL ORGANIZATION OF THE TRANSPORT LAYER

3.1.1 Low-level services architecture

Transport is the highest part of the ***low-level services*** which also include the network layer, the data link layer and the physical layer (Figure 3.1). The transport layer provides the transport services for the immediately higher layer which is the session layer in the OSI model. It uses the services of the network layer which is immediately below it.

It has been seen that the principal role of the transport layer is reliable and efficient transport of data between processes. The first question which arises is that of the type of service which must be offered to the higher layer. For many

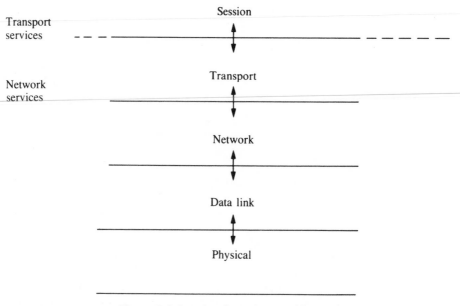

Figure 3.1 Low-level services architecture

applications, the transport layer must offer a *service with connection* which permits reliable information transfer to be guaranteed, together with sequencing and flow control. This can correspond to a *stream oriented service* where only the start and the end of the stream are delimited. In contrast, a service with connection can be organized to permit sequential transfer of *records* (*record oriented service*) which are often called *messages* or sometimes, more precisely, *letters* to distinguish them from the data units used in other layers.

The service with connection is well suited to most data processing applications and is by far the most frequently used. However, this service, which corresponds to a virtual circuit, is not well suited to the *transfer of individual messages* (*transaction oriented service*) since establishment and clearing of the connection are too time consuming for transmission of a single useful message. This leads to using a *connectionless transport service* which operates on the *datagram* principle. The connectionless service is clearly much simpler to realize than the service with connection, since it does not include connection monitoring and it performs neither flow control nor sequencing. The service is thus much less sophisticated than in the case of a connection, but the transfer times are much shorter and the transmission media are more efficiently used. The service without connection is useful for real time applications such as *data acquisition*.

Conventional applications often correspond to *dialogues* between pairs of correspondents with a hierarchical or hybrid hierarchical structure which organizes communication between one or more computers and user terminal stations. This led quite naturally to the development of networks and communication protocols which are suited to this environment and are based on services with connection. In the case of local area networks, the situation is very different since the network interconnecting the stations can be considered conceptually as a multipoint non-hierarchical data link which permits communication between any pair of stations, with all stations listening continuously. In this environment, messages can be **broadcast** from one station to all other stations connected to the network or **multicast** from one station to several other stations connected to the network. This type of exchange is not well suited to the use of a transport service with connection, since it makes acknowledgement of messages very difficult. In such a context, it is appropriate to use connectionless services.

It can be seen that the type and properties of the physical layer, the data link layer and the network layer depend very much on the system type, and the services expected of the transport layer can themselves be very different according to the intended application. This raises the problem of distributing functions among the various layers and of the degree of sophistication required in the transport layer. In the context of a conventional system, the data links are provided by a protocol with connection, for example HDLC, so as to correct transmission errors almost perfectly and to provide sequential delivery of frames. This arrangement is made necessary in practice by the fact that long-haul lines are of poor quality and that it is better to correct errors at the data link layer rather than let them propagate to the higher layers where error correction is much more difficult. In this environment, it is natural to add an X.25 network layer to the data link layer; this provides the transport layer with a practically error-free virtual circuit service and packet sequencing (Figure 3.2). The network service, in this case, is very similar to a transport service, since reliable end-to-end transport is provided for normal operation, and data is lost only in the case of network faults which are signalled to the transport layer. In this case, the transport layer can be very simple since its main role reduces to the correction of errors caused by failures signalled by the network. In addition to this function, the transport layer contains the functions which are necessary to establish and

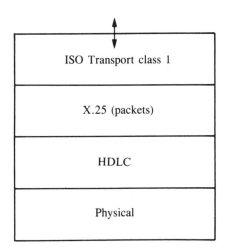

Figure 3.2 Example of the access service structure in a conventional system

clear the connection, together with the segmentation and reassembly functions between messages and X.25 packets, in order to take account of the size limits which are imposed on the latter. With the OSI model, these functions are provided by the ISO class 1 transport protocol (Figure 3.2).

For some applications, such as text transmission, the residual error rate of X.25 network connections can be considered to be acceptable, even when signalled failures are taken into account. This is particularly the case for text transmission with the *teletext service* which can be considered as an improved version of the *telex service*. The transport layer no longer needs to include a recovery function in case of network failures and it reduces to the functions of establishing/clearing the connection and fragmentation/reassembly.

In the case of a local area network, the error rate on the data circuit is very low. Hence it is logical to use a very simple data link protocol which operates without connection and without error correction. This arrangement is all the more natural since communication in this case often reduces to the exchange of a few messages between users. This leads to using a data link protocol such as LLC type 1 which is being standardized for local area networks and reduces essentially to the 'best effort' transmission of unnumbered frames (Figure 3.3). A local area network often reduces conceptually to a multipoint line to which all stations are connected. Under these connections, there is no network layer and the transport layer has direct access to the data link layer; this conforms to the OSI model which makes provision for the possibility of skipping a layer. When the user wishes to operate the local area network in connection mode and with good reliability, it is necessary to use a very sophisticated transport layer which provides residual error correction, since these have not been corrected by the lower layers. In the context of the OSI model, this can be realized with an ISO transport protocol called class 4 (Figure 3.3). This transport protocol could also have been used in the case of the preceding example (Figure 3.2), but it would have been an overkill for this application.

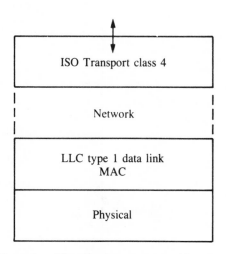

Figure 3.3 Example of the access structure with a local network

Table 3.1 Network types.

Type	Network connection characteristics	Subtype	Network service characteristics
A	Acceptable residual error rate Acceptable rate of signalled failures	A1	Reliable Sequencing Any packet size
		A2	Reliable Non-sequencing Any packet size
		A3	Reliable Non-sequencing Maximum packet size
B	Acceptable residual error rate Unacceptable rate of signalled failures		
C	Unacceptable residual error rate		

It can be seen that, for the same service provided by the transport layer, the functions which it performs depend on the quality of service provided by the network layer. It is clear that the nature of the services provided by the transport layer can also vary. The ISO has thus been led to define three types of network service which are designated by the letters A, B and C whose characteristics are indicated in Table 3.1. It is also convenient to divide type A into three categories which correspond to the different requirements in the transport layer [3.4]. This classification is useful for finding the transport connection class which must be used to provide a satisfactory service. In this context, a *residual error* corresponds to a protocol data unit which is lost, duplicated or corrupted by errors. *Signalled failures* correspond to faults which arise on the network and have caused irrecoverable errors at network level but are signalled to the transport layer. Type A networks provide a reliable service. Hence they allow the use of the simple transport layer.

In the following part of this chapter, the transport layer will be presented essentially in the context of a service with connection. However, the main features of the ISO connectionless transport layer will be indicated at the end of the chapter.

3.1.2 Services provided by the transport layer

The transport layer provides the upper layer, which is the session layer in the OSI model, with a range of services which ensure reliable and efficient end-to-

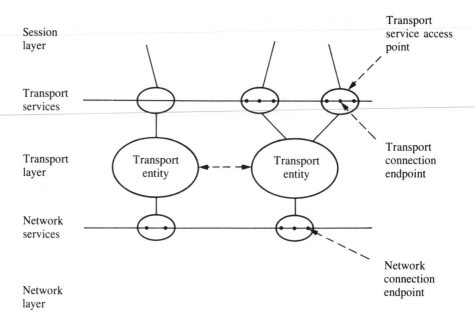

Figure 3.4 Transport layer

end data transport, with relative independence vis-à-vis the communication media. These services are provided at the interface between the transport layer and the upper layer by **transport entities** (Figure 3.4). More precisely, the main services which may be provided by the transport layer are as follows:

(a) Data transfer
(b) Service type
(c) Quality of service
(d) User interface
(e) Connection control
(f) Expedited data transfer
(g) Status reporting
(h) Security

Data transfer is clearly the basic transport layer function. The transfer must be *transparent* and is usually **full duplex**. Some systems offer the option of a **half duplex** or **simplex** transport service in order better to meet the requirements of some users.

It has been seen that the transport layer usually offers a *service with connection*, but it can also be designed with a *connectionless* service for applications with transactions of the request–response type, or for real-time applications relating to data acquisition and distribution.

The transport layer must provide *reliable and efficient* data transport. These concepts of reliability and efficiency can, to some extent, be defined a priori and taken into account when the transport layer is designed and during system installation. The acceptable reliability level and the choice of the best possible trade-off between transfer time and communication cost depend greatly on the application. On the other hand, the communication characteristics of the network can vary with time and the transport layer can affect the network support services, for example by using a larger flow control window to increase the virtual circuit throughput or by dividing a transport connection into several virtual circuits in order to increase throughput and reliability.

The transport layer thus plays the role of a bridge between the network layer and the session layer to match the service quality to the user requirements. In the case of service with connection, the **quality of service** required for communication is specified by the session layer at the time of establishing the transport connection. To do this, the transport connection request primitive contains a list of quality of service parameters which specify the levels of required service quality and relate for instance to the following points:

(a) Acceptable residual error rate
(b) Mean required transit delay
(c) Mean required throughput
(d) Priority

Following this request, the transport layer acts on the network connection parameters to obtain the characteristics which approach the user requirements as closely as possible. This is generally not sufficient and the transport layer must take appropriate measures in order to provide the required service quality. This can be realized, for example, by splitting the transport connection into several network connections to increase throughput and reliability, or by incorporating error checking in the messages so as to reduce the residual error rate. One very important action which is taken at this level is conversion of messages into packets whose size is chosen to satisfy best the specified criteria of transfer time, rate and cost of transmission. This corresponds to the operations of **segmentation**, **concatenation** and **reassembly** of data units.

The **user interface** defines the interaction between the user of the transport services and the transport entities. In the context of the OSI model, this interface is specified by the service primitives which provide a conceptual view of it. The OSI model does not specify the manner in which the user interface is implemented. In practice, the user interface can be implemented, for example, with call procedures with data and parameter passing by mailbox or direct memory access between the host computer and the front-end processor executing the transport entity.

Connection control establishes, maintains and releases the connections. The establishment and release procedures must be symmetrical with the facility for each user to initiate the operations. Releasing the connection can be performed in an *orderly* manner (*orderly connection release*) when all the data units in transit are correctly delivered before effective release. In contrast, *abrupt connection release* can involve loss of the last data units in transit on the network.

The transport layer can provide an **expedited data transfer** service which permits users to exchange data routed with high priority with respect to normal data. These expedited data units must necessarily be short and they can be used, for example, to signal an interrupt to a remote user or to transmit an alarm. The expedited data transfer service must not be confused with the priority service for normal data which is negotiated at the time of establishing the connection.

Status reporting permits the user of the transport layer to be notified of the parameters and operating conditions relating to transport which can, for example, relate to the following points:

(a) Protocol class in operation
(b) Network and transport addresses
(c) Connection characteristics – throughput and mean transit delay
(d) Quality of service degradation
(e) Current timer values

Security of transport connections concerns the precautions taken to prevent any unauthorized monitoring or modification of information originating from a transport service user. It is logical to locate protection mechanisms in the transport layer since this permits all end-to-end information to be taken into account [3.10]. Furthermore, the transport layer contains error recovery protocols which are very sophisticated and to which it is relatively easy to add security mechanisms. Possible intrusions on a transport connection can be classified as passive or active attacks. A *passive attack*, which is also called *unauthorized passive monitoring*, involves an intruder listening to the data units exchanged on the transport connections. Even if the data is coded at the presentation layer level, analysis of the headers discloses the identity and address of the correspondents, the frequency of transmissions and the length of the data units. This *traffic analysis* provides the intruder with a picture of the traffic which enables him to obtain much information on the nature of the data exchanges.

In the case of an *active attack*, the intruder acts directly on the messages exchanged on the transport connection by modification, replay, addition or deletion of data units. He can, for example, attempt sabotage by replaying a previously recorded sequence of data units at an opportune moment or attempting to establish a connection under a false identity.

When the transport layer contains protection mechanisms, the corresponding services can be requested by the user as parameters of the connection request. The OSI model provides the four following possibilities:

(a) No protection
(b) Protection against passive monitoring
(c) Protection against modification, replay, addition and deletion
(d) Protection against both types of attack

3.1.3 Functions within the transport layer

The functions of the transport layer follow from the services offered by this layer and the services offered by the network layer. The main functions are as follows:

(a) Establishment and release of transport connections
(b) Transfer of normal data units

(c) Transfer of expedited data units
(d) Mapping transport addresses onto network addresses
(e) End-to-end sequence control on individual connections
(f) End-to-end flow control on individual connections
(g) End-to-end error detection and overall quality of service monitoring
(h) End-to-end error recovery
(i) End-to-end multiplexing of transport connections onto network connections
(j) End-to-end segmenting, blocking and concatenation
(k) Supervision

The transport layer may also contain other functions, for example **accounting**, **encryption** and **temporary release of network connections**.

In the remainder of this section, the main functions of the transport layer will be presented without considering the details of protocol realization which will be examined in the following section.

3.1.4 Addressing

The principle of layer independence leads to a decoupling between network addresses and transport addresses (Figure 3.5). More precisely, connecting two transport layer users implies the use of four different identifier types. Firstly there must be a **user identifier** to distinguish a particular user from all possible users. It is then necessary to identify the **network address** which corresponds to the user transport station (*transport/network address*). The transport station can itself consist of several entities each of which corresponds to a different type of transport protocol, with, for example, the facility of operating a simple protocol to satisfy an undemanding user in the case of a high quality network and the facility of using a very advanced protocol to provide a high quality service with a poor network. This can lead to using a **transport protocol identifier**. Finally, it is possible to multiplex several transport connections onto a single network

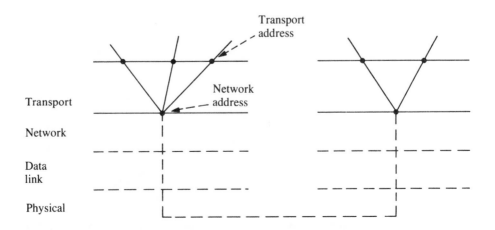

Figure 3.5 Mapping of network addresses to transport addresses

connection leading to the introduction of a ***transport connection identifier*** which is a logical connection number called ***reference*** in the case of ISO transport protocols.

With the OSI model, it has been seen (Vol. I, Chapter 1) that an (*N*)-service access point can be connected to only one (*N* + 1) entity at a given time. It is thus possible to identify unambiguously the user of a transport service by the address of one of the access points to the transport service to which it is connected. If there is only one transport address per network address (Figure 3.6), there is a one-to-one mapping between the transport and network addresses and the two address types may be identical. If a transport address is likely to map to only a single network address, which corresponds to the special case of Figure 3.6 and to the more general case of Figure 3.7, addressing can be organized in a *hierarchical* manner by concatenating the network address A with the address of the transport access point which is used, *b* in this case, to form the transport address *A.b*. In this case, the transport layer address mapping function becomes very simple since it is sufficient to extract the network address from the called transport address provided by the session entity to determine the called network address of the packets. If several network addresses can correspond to a single transport address, hierarchical address organization is no longer possible and mapping of transport and network addresses must be provided by the transport layer using a *table* which is actually a ***directory*** (Figure 3.8).

The transport address identifies the entity which uses the transport services but it is not sufficient to define the user process. In the context of the OSI model, each entity can establish several distinct connections with one or more remote transport entities (Figure 3.9).

At the time of establishing the connection, the transport layer assigns a number called the *reference* to the new connection. This reference is unique in the

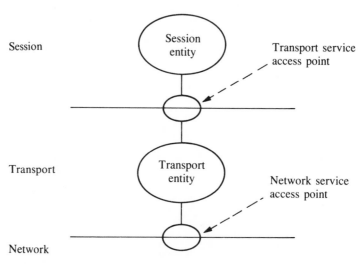

Figure 3.6 Example of a one-to-one mapping between transport addresses and network addresses

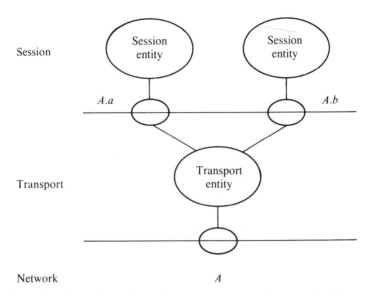

Figure 3.7 Example of hierarchical organization of transport addresses

transport layer and it has a lifetime equal to that of the connection and a purely local significance. It can be seen that the *reference* plays a double role; it permits one connection to be distinguished from another at the time of establishment by providing an identifying suffix at the connection endpoint of the transport access point, for example with the address *A.b.REF1* in Figure 3.9. After establishment of the connection, data units are exchanged with reduced headers which contain only the destination reference as an address. Using a mechanism similar to that

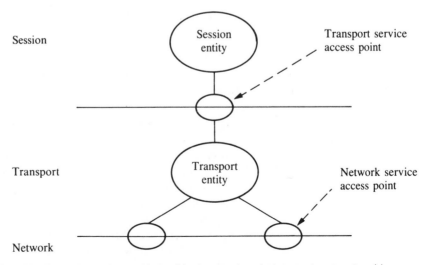

Figure 3.8 Example of non-hierarchical mapping between transport addresses and network addresses

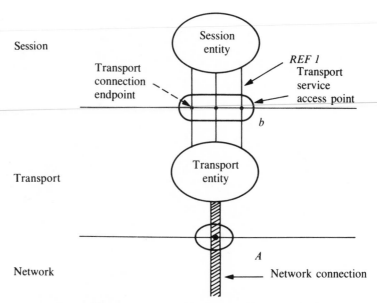

Figure 3.9 Identification of connections

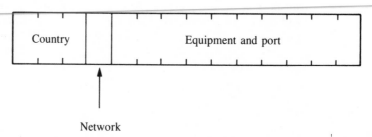

Figure 3.10 Structure of addresses for public teleprocessing networks. Recommendation X.121

described for the X.25 protocol, the transport layers use the mapping which they have established initially to deduce the addresses where the data must be sent from the reference.

So far, the problem of addressing has been presented in a relatively abstract manner, following the OSI model. More simply, addressing the user process amounts to identifying the network used for communication, the equipment in which the process resides and the process itself in the equipment. This leads quite naturally to the use of *hierarchical addressing* where the network, equipment and process addresses are concatenated. With this approach, whose principle has been seen with the OSI model, each type of address is local to the environment in which it is used; the address of equipment is local to the network to which it is connected. Similarly, the address of a process is local to the equipment or logical environment in which it is operated. Hierarchical addressing is modelled on the structure of the system; this greatly simplifies message routing since the address of each intermediate step is contained in the total address. This technique has been recommended in CCITT Recommendation X.121 for public teleprocessing networks with the addressing structure shown in Figure 3.10. The addresses contain 14 decimal digits which are divided into 3 digits for the country in which the network is located, 1 digit for the network within a country and 10 digits for the address of the equipment and the *port* to which the process is connected. The latter field of 10 digits can itself be divided into two subfields of which one is dedicated to the equipment and the other to the port.

Hierarchical addressing also has the advantage of permitting easy inclusion of *abbreviated dialling*. As each subaddress is local to the environment to which it corresponds, local communication can be performed using local addresses without reference to higher-order addresses. In this way, for example, with Recommendation X.121, communications within a country may be routed without reference to the country number. There are numerous different addressing techniques and only **global addressing** (*flat addressing*) will be presented here, which has advantages and disadvantages opposite to hierarchical addressing. With this approach, each user is identified by a global address which is unique to the system environment. It can be seen that this addressing structure is completely independent of the system structure, and this has the advantage of not involving a change of address when a subscriber moves to another location on the network. Also, the users do not have any problem in finding how to access a particular *generic service* since they have a global address in their directory.

In contrast to these advantages, global addressing seriously complicates routing and requires address management to be provided by a unique global agency for the whole system which leads to an awkward and rigid approach. For local area networks, a global approach to addressing is envisaged with 48-bit addresses which are specific to each equipment and whose assignment is managed universally. This permits a unique address for each equipment regardless of the network to which it is connected at a given time.

Some systems are based on a compromise, with essentially hierarchical addressing but with some port addresses which are global for the whole network; this permits easy calling of general purpose services (*well-known generic service address*). For teleprocessing this is the counterpart of public service telephone numbers such as the fire brigade and the police which are the same everywhere within a country.

To request establishment of a transport connection, the user process must indicate the destination address of the message which clearly assumes that it is known. In general, a process seeks to connect with a distant process or service of which it knows the name rather than the address. Hence there is a problem in matching the names and addresses, which is of the same type as that which is solved by the telephone directory for telephone users. It is solved in a similar manner by establishing *directories* which are maintained by directory management services residing in the application layer. These services can reside in the same equipment as the corresponding transport layer or alternatively in remote equipment. In the latter case, establishment of a connection can be preceded by a previous connection to a directory management service located at a global address, in order to determine the address of the called user.

3.1.5 Establishment and clearing of transport connections

The transport connection establishment phase enables correspondents to verify their mutual existence; it includes negotiation of optional parameters together with allocation of the resources required by the connection. The main functions which may be performed during the establishment phase are as follows:

(a) Selection of an appropriate network connection
(b) Decisions concerning possible multiplexing or splitting
(c) Determination of the optimum length of transport protocol data units
(d) Choice of functions which must be operational at the start of the data transfer phase, for example for error detection and recovery
(e) Limited data transfer associated with initialization
(f) Mapping of transport and network addresses
(g) Connection identification

The last two functions concern addressing. The principle of this has been mentioned in the previous section; it contributes to ensuring the independence of the transport and network layers.

The connection can be established on the initiative of any user. As the essential role of the transport layer is reliable and efficient end-to-end data transport, the establishment phase is used to negotiate the communication parameters in such a way as to ensure, as far as possible, the service quality required by the user. At the time of the establishment phase, the transport layer must decide whether the transport connection should be split into several network connections or multiplexed with other transport connections onto the same network connection. Similarly, the choice of optimum transport protocol data unit size is made during the establishment phase, which may cause the transport layer to decide to segment or concatenate messages on this connection.

In terms of the reliability required by the user and the network service quality, the transport layer selects the most appropriate network connection and sets up the error detection and recovery mechanisms which are necessary to provide the required quality of service. These error detection and recovery mechanisms may not be required if the service quality of the network is deemed to be sufficient with respect to the service quality required of the transport service. In the other case, the transport layer can summon up a message-checking system to detect lost or duplicated messages. Initialization of sequence numbers and selection of window size are performed during connection establishment. In the extreme case where the user requests a very reliable service with a poor quality network, the transport layer operates an error detecting code at message level.

During the establishment phase, the transport layer can also transfer some user data, for example to transmit a password. The transport layer can also negotiate the provision of an expedited data transfer service with the user.

It can thus be seen that the functions likely to be provided by the transport layer on a connection vary greatly with the available network services and the quality of service requested by the user. This would lead to extremely complex negotiation at establishment time if no order were imposed. In connection with the services offered by the network, this leads to classification of networks into well-defined *types*, for example with the categories shown in Table 3.1. Similarly, the functions performed by the transport layer can be grouped into ***protocol classes***. In the case of the ISO transport layer, there are five protocol classes which are numbered 0 to 4. The class numbering corresponds to an approximately hierarchical organization where a class of number N executes almost all the functions of the class of number $N - 1$ plus some additional functions. On establishment of the connection, the negotiation thus reduces to choosing the protocol class in terms of the network type and the required quality of service. The features of the five ISO/CCITT transport protocol classes will be presented in the section devoted to the CCITT X.214 and X.224 connection mode transport layer.

Connection clearing can be performed on the initative of any correspondent. It can include the following functions:

(a) Notification of the reason for clearing
(b) Identification of the cleared connection
(c) Data transfer

If there can never be more than one transport connection linked to a user, clearing of the connection can be performed *implicitly* by clearing the connection of the underlying network. With this approach, no data units are exchanged between the distant transport layers to provide clearing. The clearing request expressed by a transport service user at one end leads directly to clearing of the corresponding network connection; this causes clearing of the network and then transport connections at the other end. This method does not ensure an ordered clearing in the sense that the data units still in transit on the network following an unusual delay can be lost at the time of clearing. Users of the transport layer must, therefore, ensure that all messages have been received before ordering clearing of the transport connection.

When several transport connections are multiplexed on to the same network connection, the preceding method is no longer applicable since the network connection must be maintained until the last transport connection which uses it is cleared. Clearing of the transport connection in this case implies sending a data disconnection unit to the remote transport entity to cause clearing at the other end of the connection. Clearing can again be performed in an abrupt manner with possible loss of data units in transit at the time of disconnection, or in an ordered manner with a guarantee of correct delivery of all data units.

3.1.6 Data transfer

The transport layer transfers data between two session entities using full duplex transmission of ***transport protocol data units*** (***TPDU***). According to the class of protocol selected on establishment, the transport layer can operate some of the following functions:

(a) Segmentation
(b) Grouping

(c) Concatenation
(d) Multiplexing or splitting
(e) Delimiting of transport service data units
(f) Sequencing
(g) Flow control
(h) Error detection
(i) Error recovery
(j) Identification of transport connections
(k) Expedited data transfer

Once communication is established, the **transport service data units** (**TSDU**) are transferred in the form of TPDU which contain only the connection number, called the *reference* (Figure 3.11), as an address indication. The TPDU contains a header called **transport protocol control information** (**TPCI**) which might contain, for example, the reference, the TPDU type indication and a sequence number.

The TSDU are, in fact, *messages* corresponding to the session entity whose size is dictated by the application and can in principle be of any length. The TPDU correspond to *packets* which are routed by the network and, as has been seen several times, they must have a size suited to the communication media to permit both fast and economical transmission. Also, the network generally imposes a maximum packet size to satisfy technical requirements such as the limited size of the buffers used in the network nodes. The network layer must, therefore, segment long messages into several packets before transmission on the network, or group several short messages into one longer packet in order to utilize the network with a packet size which minimizes transfer time (Figure 3.12).

More precisely, **segmenting** consists of separating one TSDU into several TPDU. At the other end the TPDU must clearly be subjected to the inverse operation of **reassembly** to reconstruct the original TSDU (Figure 3.13). To make the reassembly operation possible, it is clearly required that the various TPDU forming the same TSDU should be received in sequence and each TPDU should

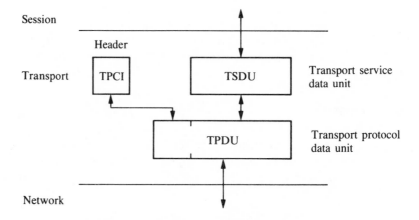

Figure 3.11 Mapping between data transport units

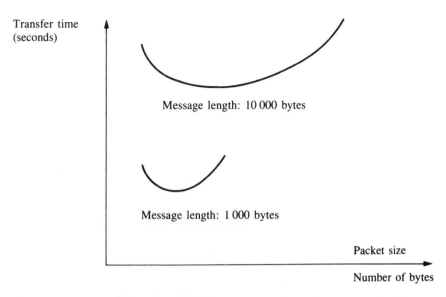

Figure 3.12 Typical relationship between transit time and packet size in a packet switching network

contain a parameter indicating whether or not this packet is the last of the message. As each TSDU is a whole message at the session/transport interface, delivery of a TSDU to the destination session entity must not be made until after complete reassembly. Segmentation and reassembly are two fundamental functions of the transport layer, since it is these which permit the network to operate by packet switching instead of message switching due to decoupling between message size and packet size.

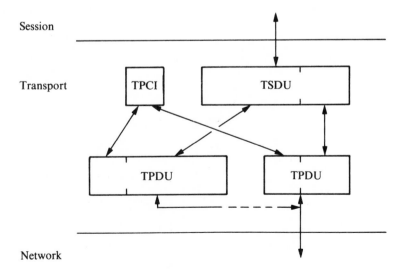

Figure 3.13 Segmentation and reassembly

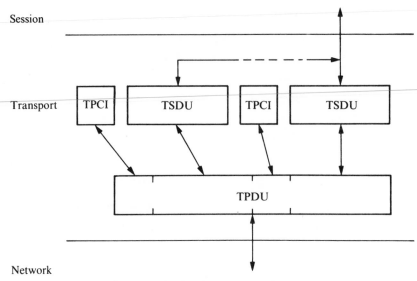

Figure 3.14 Blocking and deblocking

Blocking, which is not often used at the transport level, consists of assembling several TSDU into a single TPDU (Figure 3.14). It is used to assemble several short messages into a single packet whose size is close to the optimum. The inverse operation is *deblocking*.

Blocking and segmentation concern only the data units transmitted on the same connection. The transport protocol can also perform *concatenation* in which the TPDU belonging to various transport connections are assembled into the same *network service data unit* (*NSDU*) transmitted on a network connection (Figure 3.15). This function appears as a multiplexing of several transport connections onto a network connection, but it is different since, in this case, there is no

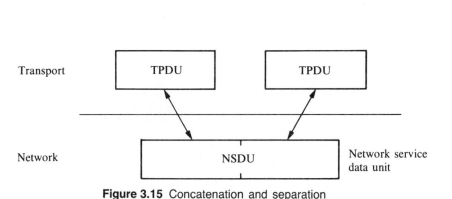

Figure 3.15 Concatenation and separation

permanent association between the network connection and the transport connections. Also, the TPDUs corresponding to the various transport connections are transmitted in the form of distinct NSDU in the case of multiplexing. Concatenation is more advantageous than blocking when the traffic on the transport connections consists of short messages interspersed with long idle periods. In this case, the waiting required to group several TSDU into a single packet introduces an excessive delay and it is generally possible to obtain a packet of the same size with a shorter waiting delay by concatenating the TPDUs belonging to different transport connections.

3.1.7 Multiplexing and splitting

Multiplexing consists of mapping several transport connections onto a single network connection (Figure 3.9). It can only be performed end-to-end; this corresponds to the case where several transport connections are operated in parallel on the same network connection which links two pieces of equipment. It can be seen immediately that multiplexing requires flow control at the transport level, since flow control performed by the network acts on the multiplexed traffic and not on each individual transport connection.

Multiplexing is mandatory when several transport connections must be established through a network which can provide only a single virtual circuit between two users. In practice, networks are almost always able to establish a large number of distinct virtual circuits between two users and it has been seen that this is the case for X.25 networks. Under these conditions, the use of multiplexing at the transport level can be questioned since the network can assign a distinct virtual circuit to each transport connection. The reply to this question is that establishment and maintenance of a virtual circuit takes up resources at the network level since the latter must create a virtual circuit descriptor and allocate buffers to the circuit. In the case of a public network, this is reflected in the invoicing which is partly related to the number of virtual circuits used. Under these conditions, it is advantageous for the user to multiplex several transport connections on to the same virtual circuit, up to a particular through-put limit which is determined by the network flow control window. Also, multiplexing of transport connections cannot be avoided when several transport stations access the network at the same point. The stations can then be in conflict by accidentally choosing the same virtual circuit number if they do not operate on multiplexed connections.

Splitting is the function which makes several network connections correspond to one transport connection (Figure 3.16). The inverse function is *recombination*. Splitting is used by the transport layer to increase the throughput at network level by operating several virtual circuits in parallel. With this approach, the TPDUs to be transmitted are distributed cyclically on the N network connections. As the throughput on each virtual circuit is limited by the window size, the maximum throughput on the set of N virtual circuits is multiplied by N on condition that it does not exceed the maximum capacity of the data link. Splitting can also be used to improve the availability of the transport connection. It is

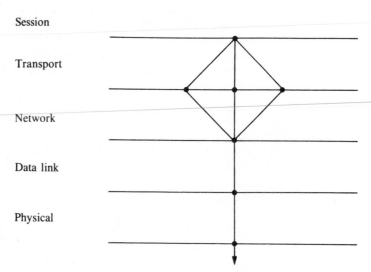

Figure 3.16 Splitting at the transport level with multiplexing at the network level

then necessary that each virtual circuit on which the transport connection is split corresponds to a distinct path at the data link and physical layer levels so that a break at any level cannot interrupt communication (Figure 3.17). This means taking particular precautions in order to avoid, for example, two redundant lines using the same conduit or the same cable, which could involve a simultaneous break of the two lines in the case of an incident.

3.1.8 Sequencing, flow control and error checking

When the network guarantees sequential delivery of packets which are entrusted to it, the transport layer does not need to operate a sequencing function.

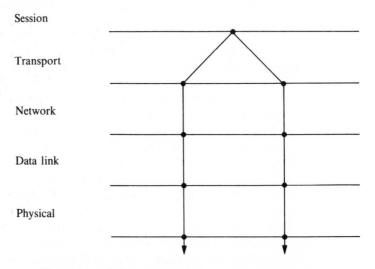

Figure 3.17 Splitting with separate routes between users

Otherwise, *sequencing* is ensured by numbering the TPDUs in sequence according to a principle similar to that described for data link protocols and X.25.

Sequence numbering of the TPDUs can also be used to detect lost or duplicated packets and thus to provide error checking at the transport protocol level. Transmission errors are normally corrected in the data link layer so it is, in principle, superfluous to add redundancy to the TPDUs with an error-detecting code. This technique is used only when the transport station uses a poor quality network.

It has been seen that it is required to use *flow control* in the transport layer if multiplexing is used. At the transport level, flow control can use three types of interactions:

(a) Interactions at the session/transport interface
(b) Interactions at the transport/network interface
(c) Interactions between distant transport entities

Interactions at the session/transport and transport/network interfaces have a purely local direct effect but can cause a chain of events which propagates by **back pressure** to a remote entity. In this way, for example, the session/transport interface can be organized in such a way that the session entity can reject a message which is submitted to it by the local transport entity when its receive buffer is congested. This will cause retransmission at the other end of the lost message and hence a traffic increase which in turn can cause congestion in the lower layers.

If flow control is performed at the transport/network interface, any throughput reduction imposed on a network virtual circuit by a transport connection is reflected on the other transport connections multiplexed on to the same network circuit.

Flow control at the transport level can thus be performed in a satisfactory manner only by cooperation between the distant transport entities at the transport protocol level. This leads quite naturally to the concept of a flow control mechanism similar to that which is used with data link and network protocols. In this case, the rate of TPDU transmission is controlled by the acknowledgements, for instance with a sliding window protocol.

It can be seen that even when the network is perfectly reliable, flow control requires TPDU sequence numbering and an acknowledging mechanism in the transport layer. It will be seen that this mechanism is distinctly more complicated than those which are used in the network and data link layers.

3.2 ELEMENTARY TRANSPORT PROTOCOL MECHANISMS

3.2.1 Problems arising in the design of a transport protocol

The design of a transport protocol can seem to be very similar to that of the network and data link protocols. In fact, it will be seen in this section that there are important differences which arise for two reasons. The first is associated with the fact that the transport layer is the last service layer which controls data transfer between users. The transport layer must, therefore, resolve all problems which have been left aside by the lower layers and it must be even more sophisticated if the lower layers are less advanced. In this context the usefulness will be seen of solving problems at the lowest level where they can be controlled. This will be illustrated by the fact that transport protocols in certain cases can only imperfectly correct errors or failures left over by the lower layers.

The second important difference between transport protocols and network and data link protocols relates to the fact that the network has memory and that the transfer times between transport entities are long, variable and unpredictable. This requires taking account of the extreme case of a rogue packet which remains in the network for an abnormally long time; this causes it to arrive out of sequence with other packets sent before it or even after clearing the connection.

The transport layer is as simple as possible in the case of a reliable network with a negligible rate of residual errors and signalled failures; this corresponds to a type A network of Table 3.1. If, in addition, the network is of subtype A1, with a sequencing service and no packet size limitation, the transport protocol functions reduce to establishment and clearing of the connection, addressing, flow control and multiplexing. Notice that it is possible to skip the transport layer if there is no multiplexing of transport connections, since a perfectly reliable network can be assumed to ensure end-to-end transport and flow control can be provided by the network layer. Even in this case, it can be useful to use a rudimentary transport layer which performs message segmentation in order to utilize the network more efficiently. If several transport connections must be multiplexed onto one network connection, the transport layer becomes mandatory since it must be able to establish and clear distinct transport connections. The transport layer must also contain a mapping function for transport and network addresses, together with a flow control function on each transport connection.

When the network can still be considered to be reliable but when it no longer ensures packet sequencing, which corresponds to subtype A2, the transport layer must ensure packet sequencing in addition to the functions corresponding to subtype A1. This poses additional important problems particularly for processing rogue packets at the time of establishing and clearing connections. If the network also imposes a maximum packet size, which corresponds to subtype A3, the transport layer is compelled to contain message sequencing and delimiting functions so that it can segment messages whose size exceeds the maximum packet length.

In the case of a type B network, packets are still routed without error, without loss and without duplication in normal operation. However, in this case the network can have operational failures which lead to the loss of some packets. These failures are signalled to the transport layer so that this can undertake a recovery procedure. A situation of this kind is found with an X.25 type network where a RESET command can cause the loss of packets in transit. Even when there is no multiplexing, a transport layer is required to ensure reliable end-to-end data transfer. On the sender side, the transport layer must save the transmitted TPDU until correct delivery in order to be able to provide recovery in case of a signalled failure.

The most difficult problem concerns the design of a transport layer capable of operating with a type C network which has an unacceptable residual error rate for its normal operation. In this case, the TPDU can be damaged or lost during transfer and the transport layer must be able to check the integrity of the received TPDU by means of an error-detecting code. The TPDUs thus contain an error check word and correction of erroneous TPDUs is ensured by error detection, acknowledgement and retransmission. Retransmission of lost TPDUs is initiated by an acknowledgement timer.

3.2.2 Transport connection establishment mechanisms

Establishment of a transport connection starts on the initiative of one of the transport layer users by sending a T-CONNECT request primitive (TCONreq) to a transport entity. The primitive contains, among others, the source and destination addresses together with the quality of service parameters (Figure 3.18). On receipt of this primitive, the local transport entity assigns a source *reference* to the connection and sends a *connection request (CR)* TPDU to the remote transport entity after having selected one or more network connections which it maps onto the transport connection (Figure 3.19). The CR TPDU serves to signal the connection request to the remote transport entity and to negotiate the connection parameters. Its arrival causes assignment of a destination *reference* to the connection and transmission of a T-CONNECT indication primitive (TCONind) which notifies the connection request to the remote user and passes the requested parameters to it. If the request is accepted by the remote user, the latter responds with a T-CONNECT response primitive (TCONrsp) which permits the corresponding transport entity to return a *connection confirm (CC)*

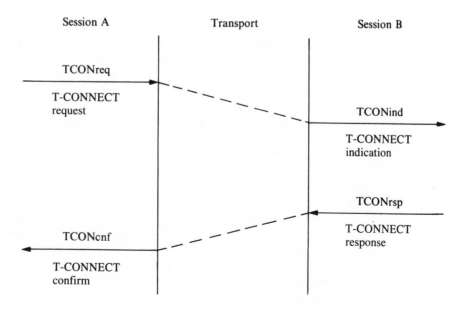

Figure 3.18 Transport connection establishment

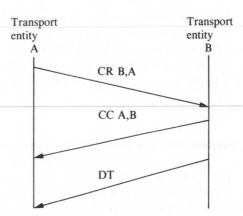

Figure 3.19 TPDU exchanges for the successful establishment of a transport connection

TPDU to the sender. This CC TPDU contains the response to the connection parameter negotiation together with the source and destination *references* of the connection which permit the sending transport entity to map the two references. The exchange is terminated by sending a T-CONNECT confirm primitive (TCON cnf) to user A. The transport connection is now established and data exchanges can start, for example on the initiative of user B who initiates transmission of a data unit in the form of a *data* (*DT*) TPDU. Data TPDUs are usually numbered in order to permit sequencing, error detection and flow control using a mechanism similar to that which is used with HDLC and which will be described in the following sections. It will be assumed firstly that the data TPDUs contain a send sequence number $T(S)$ which is initialized to zero in each direction at the time of circuit establishment and is computed modulo N.

Satisfactory conclusion of connection establishment implies firstly that the distant user accepts the connection and above all that no hazard has occurred on the network so that the latter is perfectly reliable. When the distant user rejects connection establishment, the latter signals his rejection by responding to TCONind with a T-DISCONNECT request primitive (TDISreq). This primitive causes clearing of the connection at the remote transport entity and sending by it of a *disconnect request* (*DR*) data unit TPDU (Figure 3.20). The DR TPDU initiates clearing of the connection at the calling transport entity and sending by the latter of a T-DISCONNECT indication service primitive (TDISind) which signals the connection reject to the local user. The connection can also be rejected on the initiative of the remote transport entity if it perceives the received CR data unit as unacceptable. In this case, the remote transport entity refrains from establishing the connection and merely returns a DR data unit to cause clearing on the calling side.

When the network is subject to failures, establishment of the connection can be subject to numerous hazards. Consider firstly the case where the network can still be considered to be perfectly reliable but where perfect sequencing of TPDUs is no longer guaranteed. A first problem arises from the fact that one of the user

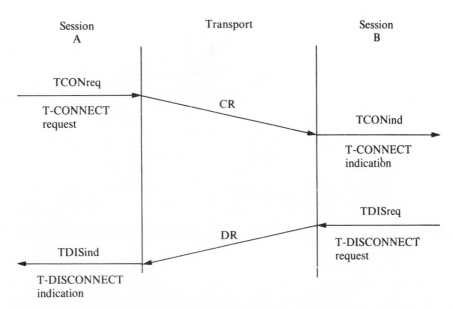

Figure 3.20 Refusal of the establishment of a transport connection by a distant user

data units DT transmitted by the remote entity may arrive at the local transport entity before *connection confirm* (*CC*) and thus before the connection is formally established (Figure 3.21). In such a case, the logical solution is to ignore the TPDU DT which arrive before confirmation of connection. This is paradoxical since it causes loss of data units although the network is perfectly reliable. Under these conditions, one could think of storing the DT TPDU until connection confirmation in order to restore the sequence when the connection is established. This solution is also not very satisfactory since it causes the transport layer to

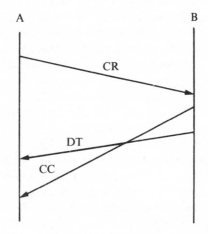

Figure 3.21 Arrival of a user data unit before connection confirmation in the case of a nonsequencing network

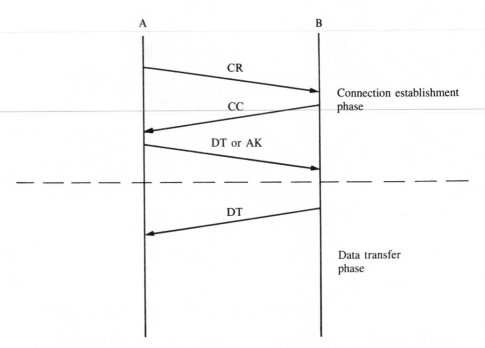

Figure 3.22 Three-way handshake procedure for transport connection establishment

operate temporarily in an ill-defined state, which can be the source of new problems. It is possible to arrive at a much more satisfactory solution by adopting a *three-way handshake procedure* where the called entity is not allowed to send a TPDU other than CC until it has received a data unit which confirms correct reception of CC by the call initiator (Figure 3.22). The transport connection establishment procedure is asymmetric in this case since the caller must confirm correct receipt of CC by sending a DT TPDU if it has data to transmit, or otherwise by sending a *data acknowledgement* data unit AK.

The three-way handshake mechanism satisfactorily resolves the problem posed by out-of-sequence transmission of the first packets by the called entity, since they cannot be sent until after complete establishment of the connection. Use of a nonsequencing network can be the origin of more subtle faults which are due to the possible survival of packets belonging to a connection even after it has been cleared. Consider the case where a connection between two entities A and B is cleared in a disorderly manner by the exchange of a DR TPDU and a DC TPDU (*disconnect confirm*) (Figure 3.23). As the network does not guarantee packet sequencing and can at times introduce considerable transfer delays, some TPDUs transmitted on the connection may have not yet arrived at the time of clearing and it is possible that they continue to circulate for a long time in the network before arriving at their destination. Assume now that entity A again opens a connection with entity B by choosing the same *reference* as for the previous connection. After initialization, a surviving data unit from the first connection can arrive at A with the *reference* of the new connection. In this case,

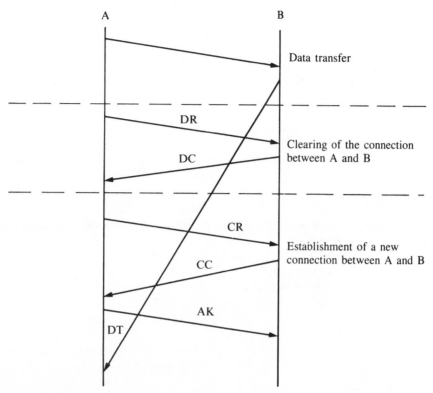

Figure 3.23 Example of protocol error on a connection caused by the survival of packets exchanged on a preceding connection

entity A has no means of detecting that this data unit belongs to the previous connection. It carries a correct destination *reference* and, as sequence numbering is modular, it can even be received in sequence with the data units belonging to the new connection. If this is not the case, the rogue data unit will not be considered to be incorrect since it is a normal state of events to receive out of sequence data units on this network.

To resolve the problems posed by the resurrection of packets belonging to a dead connection, a first possibility is to impose an ordered closure of the connection in such a way as to ensure that all packets have been correctly received before clearing. This ensures that no rogue packets occur after closure if none of them have been duplicated, which is the case with a protocol designed for a reliable non-sequencing network. In contrast, some packets could be duplicated if the network is not reliable and correct receipt of all the data units does not guarantee the elimination of surviving replicas. Moreover, ordered closure can lead in some cases to needless use of transport entity resources until the arrival of the last data units.

The problem of rogue data units can be resolved in a more satisfactory manner by transferring the solution to the network layer with a mechanism for limiting packet lifetime. This technique consists, for example, of assigning to each packet transmitted on the network a maximum number of nodes through which it can pass; this number is decremented by one count during each crossing of a node, and packets are discarded when the count is zero. With this arrangement, the maximum packet lifetime is limited and it is sufficient to impose a maximum delay between clearing and subsequent establishment of the same connection to control the problem. This delay can be a nuisance and it is preferable to use a **frozen reference** method which consists of prohibiting reuse of a connection *reference* which has just been cleared for a time at least equal to the maximum packet lifetime.

When the network is likely to introduce transmission errors, the problems which have just been mentioned become more complex, since some data units can be lost, duplicated or in error. Connection establishment can be disrupted by loss of the *connection request (CR)* or by loss of the connection confirmation acknowledgement AK or DT in the case of a three-way handshake procedure. Protection against faults is based on the use of a **retransmission timer** which is set to a guard time $T1$ greater than the round trip transfer delay of a data unit on the network. The timer is set on transmission of the data unit and the latter is retransmitted if no acknowledgement has been received on the expiry of the guard time. With this approach, the loss of a CR TPDU causes a simple retransmission without duplication at the destination entity. In contrast, loss of the *connect confirm* CC initiates retransmission of the CR TPDU which causes duplication of the latter (Figure 3.24). If the CC TPDU is only very much delayed instead of lost, there is also duplication of the CR TPDU at the source of the connection request. The behaviour of the protocol is simply to ignore the CR and CC TPDU replicas when the connection is established.

Replicas of *connection requests* can cause erroneous establishment of a new connection when they follow clearing of the previous connection after an excessive network delay. A CR TPDU of this kind is considered as a genuine connection request by the entity which receives it (Figure 3.25). This causes it to accept the request and establish the connection on the called side by transmitting a *connection*

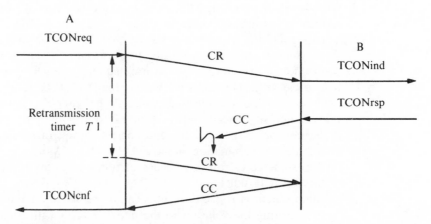

Figure 3.24 Monitoring the establishment of a connection by a retransmission timer

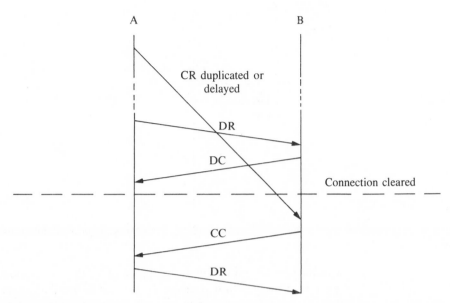

Figure 3.25 Erroneous establishment of a connection by a duplicated and delayed connection request

confirm CC. On the calling side, the *connection confirm* is then rejected since the corresponding entity has not made the connection request. If nothing is done, the connection would be open on one side and closed on the other, which is clearly unacceptable. The solution to this problem implies the destination of CC sending a *disconnect request (DR)* to the transport entity which erroneously opened the connection.

3.2.3 Transport flow control

When the network is perfectly reliable, with sequencing and without packet size limitation, data transfer at the transport level reduces simply to sending a DT TPDU user data unit following a T-DATA request primitive (TDTreq) (Figure 3.26). On the remote transport entity side, receipt of the DT data unit causes data to be sent to the destination by a T-DATA indication primitive (TDTind). The two primitives contain the data as parameters. If there is only a single user of the transport layer, the transport connection is unique and there is no ambiguity in the destination of data units. Under these conditions, the data transfer primitives need not mention the *reference* and this need not appear in the DT data unit. In contrast, when there are several transport connections, each one of them must be identified by a *reference* in the DT TPDU and the data transfer primitives.

As the network is perfectly reliable with sequencing, data transfer does not require any acknowledgement mechanism. Moreover, as the network imposes no limit on maximum packet size, messages can be transferred without segmentation on the basis of one message (TSDU) per packet (TPDU). If the network operates

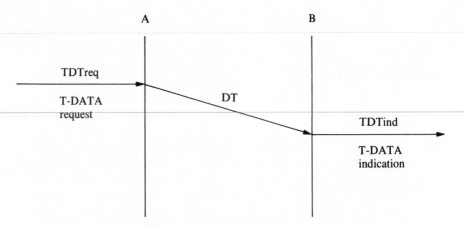

Figure 3.26 Data transfer on a reliable network

with packets whose maximum size is limited, the only modification required is to segment messages into packets. In this case, the data must not be delivered to the user of the transport layer until after the arrival of the last packet of the message which requires this to be marked with an *end of TSDU mark* (*EOT*) (Figure 3.27).

In connection with flow control, the following five solutions can be considered:

(a) No control
(b) Control by the network layer
(c) Stop and go control
(d) Sliding window with fixed credit
(e) Sliding window with variable credit

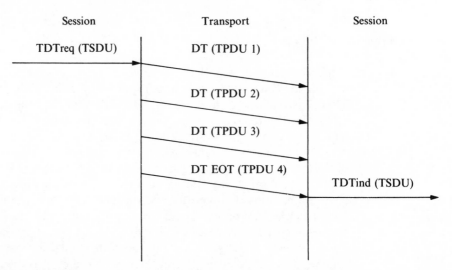

Figure 3.27 Data transfer on a reliable network and segmentation of messages into packets

In the absence of flow control, some data units must be discarded on arrival because of a lack of free buffers in the remote transport entity. This type of operation is clearly paradoxical since it is then necessary to provide a retransmission mechanism although the network is perfectly reliable. It is, therefore, generally necessary to provide a flow control function. However, as the latter consumes resources and can reduce the maximum throughput, it is often useful to offer the facility of ruling out flow control when the destination has no input traffic limitation and the throughput must be as high as possible.

It is also possible to consider transferring flow control to the network layer by controlling traffic flow at the interface between the network and transport layers. This solution can, if necessary, be used when several transport connections are not multiplexed onto the same network connection but it must not be used otherwise, since this would lead to slowing down exchanges on all the transport connections when only one of them is overloaded.

It can be seen that this usually leads to the provision of individual flow control on each transport connection. The problem is more difficult to solve in this case than in the data link and network layers since the data units can suffer delays which can be very long and whose duration is unpredictable. Because of this, transmission rate increase and decrease commands, which are sent by the destination, can very well give the sender of the data units an outdated picture of the real situation. To confirm this, consider, for example, the case of a transport layer where the receive buffers are shared among various connections. If there are several free buffers at a given time, the destination transport entity can transmit to the sender a command which authorizes increasing the throughput on the connection. It can be seen immediately that when this command takes a long time to reach the sender, it can cause it to increase the transmission rate although the receive buffers have been seized by other connections in the meantime.

Flow control is an automatic control problem which becomes more difficult to solve as the loop delay increases. To prevent instabilities, it is tempting to choose a conservative strategy which tends to set the system clearly below its maximum throughput. However, it will be seen that this approach can reduce the throughput excessively since this is essentially determined by the flow control parameters. It can be seen that flow control design implies a difficult trade-off between flow control efficiency and optimization of the network throughput.

The first possible flow control mechanism is a stop and go operation such that the destination transport entity sends commands which authorize or halt the transmission of data units by the sender. This control mechanism is similar to that which is used in some data link protocols with the XON and XOFF commands. It is very badly suited to a transport protocol, except possibly in a local environment, since during the, sometimes long, delay which occurs between a go and a halt command, the sender can transmit a number of data units greater than the number of free buffers available at the destination.

A more elaborate method of flow control consists of using a *sliding window* protocol which is similar to that used in the HDLC data link protocol. With this approach, the DT TPDU are numbered sequentially and their rate of transmission is controlled by the rate of receipt of acknowledgements. More precisely, the source and destination maintain a send sequence number $V(S)$ and a receive sequence number $V(R)$ on each connection; these are both reset to zero at the time of circuit establishment and are defined modulo M. The send sequence number $V(S)$ is the number of the next user data unit to be transmitted. During transmission of a DT TPDU, the source writes the current value of $V(S)$ into a field $T(S)$ of the DT TPDU, then increments $V(S)$ by one.

The receive sequence number $V(R)$ represents the number of the next expected DT TPDU. During receipt of a DT TPDU, the destination transport entity checks that its number $V(R)$ is equal to the number $T(S)$ of the data unit DT. If this is the case, it increments its receive sequence number by one which may permit it to return an acknowledgement AK to the sender; this contains the number $V(R)$ of the next expected DT TPDU in its $T(R)$ field.

When the protocol window is fixed at a value W, the sender who has received an acknowledgement bearing the number $T(R)$ can send the W data units DT, whose numbers are from $T(R)$ to $T(R) + W - 1$ (Figure 3.28). Transmission must then stop if a new acknowledgement has not been received. The protocol thus ensures that there are never more than W DT data units in transit on the network. To provide a better efficiency, the acknowledgements can be grouped with the rule that an AK TPDU bearing the number $T(R)$ acknowledges receipt of all DT TPDUs whose sequence numbers are less than $T(R)$.

With this approach, the receiving entity can provide flow control in a very simple manner by reducing the rate of transmission of acknowledgements, or even stopping them, which causes the sending of DT TPDUs to stop after expiry of the transmission credit represented by the window W.

In the case of a reliable network with sequencing, and when the maximum packet size is such that no segmentation is required, flow control with a fixed sliding window applies very well and often provides a service well suited to the application. The data units in this case are always correctly routed and transferred in sequence so that the acknowledgements serve only for flow control and there is no need to save a copy of the transmitted DT TPDU at the sender. On the other hand, as there is no limit on packet length, segmentation is superfluous

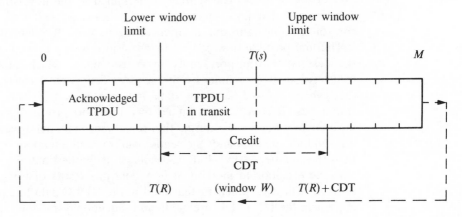

Figure 3.28 Flow control window

and each message is transmitted by a single DT TPDU which can be passed to the session layer as soon as it is received by the destination transport entity. It can be seen that only one buffer is required at the sender and one at the receiver. This implies that flow control is only necessary if it is performed for the benefit of the session layer.

Now consider the case of a reliable sequencing network, but with a limitation on maximum packet size. In this case segmentation can be used so that the sender must use a buffer capable of storing the complete message corresponding to one TSDU before segmenting it into several TPDU. Similarly, the destination must have a receive buffer capable of storing one TSDU, since it is necessary for it to reassemble the DT TPDU into a complete message before delivery to the user. The buffer size which must be provided is, as before, equal to one message on the transmission side and one message on the receive side so that flow control is justified only if it covers the requirements of the session layer.

When the network is reliable but without sequencing, the DT TPDUs cannot be delivered to the user until they have been put back into sequence by the destination transport entity, regardless of the message length. This poses a new problem which is only imperfectly solved by the sliding window protocol. The destination can return an acknowledgement only if it has received at least the DT TPDU whose sequence number immediately follows that of the last DT TPDU which it has acknowledged (Figure 3.29). If a DT TPDU is very much delayed, transmission of subsequent data units stops when W data units have

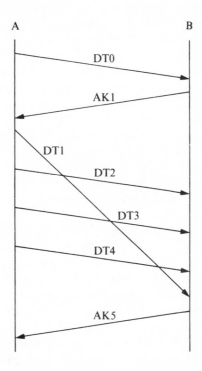

Figure 3.29 Acknowledgement delay caused by faulty sequencing

been sent, since the acknowledgement will not be able to arrive before expiry of the credit represented by the window. It can be seen that the receiving transport entity must be equipped with W receive buffers and the window W must be as large as possible to avoid temporary deadlock. These requirements are contradictory and this leads to the concept of adaptive flow control, where the window size can be adjusted dynamically.

When the network is subject to failures which are signalled to the transport layer, the data units are transmitted correctly in the normal state, but some of them can be lost following a network failure which is signalled to the transport layer by an N-RESET indication or N-DISCONNECT indication primitive. There is no need to provide an error correction mechanism in the normal state, but it is necessary to initiate a recovery procedure when a failure is signalled, since this could have led to the loss of DT data units or acknowledgements. To achieve this, the sender saves a copy of the transmitted TPDUs until it has received the corresponding acknowledgements; they may then be discarded. When a network failure is signalled to it, the destination must indicate to the sender the number of the next expected DT TPDU in order that the latter can retransmit those which are lost. The protocols which have been described so far do not include any retransmission procedure, since a DT data unit can wait for transmission until the acknowledgement which frees the transmission arrives, and receipt of an AK TPDU replica does not affect the sender.

It is, therefore, necessary to provide a retransmission mechanism based on the destination sending to the source one *reject* data unit RJ which contains the number $T(R)$ of the next expected DT TPDU; this RJ TPDU is sent after a network fault is signalled. On reception of this RJ data unit, the sender must transmit the DT TPDU whose number $T(S)$ is equal to the number $T(R)$ of the RJ TPDU, even if that corresponds to a retransmission.

With a network whose error rate is unacceptable, some TPDU can be lost or in error in the normal state without this failure being signalled by the network. In this case, the system must operate a mechanism which ensures retransmission of lost or impaired data units. This is generally realized with the help of a *retransmission timer* which is set when a DT data unit is sent and whose guard time $T1$ is greater than the sum of the transfer times of the DT TPDU and its corresponding acknowledgement AK (Figure 3.30). The sender must save a copy of each transmitted data unit until it has received the corresponding acknowledgement. If the acknowledgement has not arrived on expiry of the guard time, the sender must consider that the corresponding DT TPDU is lost and provide retransmission. This mechanism can clearly cause a duplication of the DT TPDU if the fault has affected the acknowledgement instead of the DT TPDU, but this is not important since the destination can recognize the duplicated data units by means of the sequence number $T(S)$.

It can be seen that, with an unreliable network, the acknowledgements play a triple role, since they serve for sequencing, error detection and flow control. This is the origin of a new problem which arises from the fact that acknowledgement of a DT TPDU must, in the absence of an error, arrive before expiry of the guard time. It is, therefore, no longer possible to stop the transmission of DT TPDUs by suspending the sending of acknowledgements, since this would cause

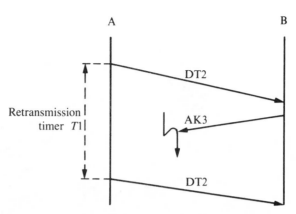

Figure 3.30 Correction of data units lost or in error

unnecessary retransmission. A first solution to this problem is to define a new data unit which acknowledges the DT TPDUs whose number is less than its number $T(R)$ but which prohibits sending of new DT TPDU until a regular acknowledgement is received. Such a data unit would play a role similar to the *receive not ready* frame RNR of the HDLC protocol. However, it is possible to operate an approach which realizes this function in a more flexible manner by modifying the acknowledgement mechanism to permit dynamic modification of window size. This technique will be described in the following section.

3.2.4 Flow control with a credit mechanism

The principle of *adaptive flow control* consists of providing an additional field in the acknowledgement data unit AK which contains a *credit* CDT representing the number of DT data units which the destination is ready to receive starting from the number $T(R)$ of the next expected packet. Receipt of this acknowledgement thus defines a sender window whose lower limit is $T(R)$ and whose upper limit is $T(R) + \text{CDT}$; this authorizes transmission of data units whose send numbers $T(S)$ are from $T(R)$ to $T(R) + \text{CDT} - 1$ (Figure 3.28). Sending data units whose number is greater than $T(R) + \text{CDT} - 1$ cannot be performed until after reception of a new acknowledgement which increases the upper limit of the window.

The adaptive control mechanism can be illustrated by the example of Figure 3.31 where entity A has an initial credit of 5 which allows it to send data units DT0 to DT4 in succession. After receipt of DT0 and DT1, entity B globally acknowledges these two data units with an AK data unit which carries the number $T(R)$ of the next expected DT TPDU, 2 in this case, together with a credit of 3. The upper window limit thus remains at 5 so that the sender must cease transmission after sending data unit DT4. The data units are subsequently acknowledged by acknowledgement AK5 which carries a credit of 0. The upper window limit remains unchanged which thus blocks the sender. After some time,

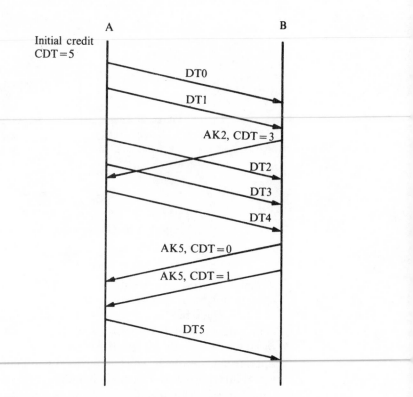

Figure 3.31 Example of flow control with credit allocation

entity B sends an acknowledgement which carries a credit of 1 in order to indicate that it has an available buffer to receive an additional DT TPDU. This acknowledgement thus unblocks entity B by permitting data unit DT5 to be transmitted.

The credit allocation mechanism permits very precise flow control, since it provides the destination with the facility of allocating a sender credit exactly equal to the number of buffers reserved for receiving on the connection. Moreover, this technique solves the problem of flow control on a low reliability network, since an acknowledgement with zero credit acknowledges receipt of despatched data units while blocking the sender of additional DT TPDUs.

Operation of the adaptive flow control procedure requires certain rules to be observed which depend on the data transfer protocols used. Consider, firstly, the case of a reliable network, for which the transport protocol does not normally use any error-detection and retransmission mechanism. If an acknowledgement is sent with $T(R)$ and CDT values such that the upper window limit $T(R)$ + CDT is less than the previous limit, it can well happen that a DT TPDU has been transmitted outside the window before the acknowledgement has arrived at the sender. This DT TPDU must be rejected since there is no free buffer to receive it and it is thus irretrievably lost as there is no retransmission mechanism. An event of this kind is represented in Figure 3.32 where the upper window limit

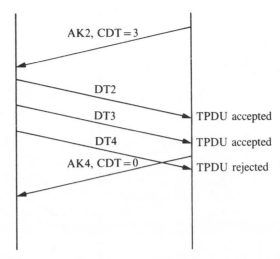

Figure 3.32 Loss of a data unit as a result of a reduction in the upper window limit

which was initially 5 is subsequently reduced to 4 by acknowledgement AK4, CDT = 0; this causes the loss of data unit DT4.

In a network with an unacceptable rate of signalled failures, the transmission protocol uses a mechanism for saving a copy of the DT TPDU until the arrival of the corresponding acknowledgements, with the possibility of retransmission upon request by an RJ data unit which acknowledges the correctly received DT data units and permits retransmission of the following ones. This procedure can be used to reduce the upper limit of the window with the help of an RJ TPDU. Reconsidering the previous example, it can be seen that the upper window limit can, in this case, be reduced by replacing the regular acknowledgement by a reject data unit RJ4, CDT = 0, which blocks the sender but prompts retransmission of DT data units whose number is equal to or greater than 4 (Figure 3.33). When a buffer is free, the destination can then unblock the sender with a regular acknowledgement, which in this case is AK4, CDT = 1. With the example of Figure 3.33, this causes retransmission of data unit DT4 which is accepted by the destination.

Now consider the case of a network with an unacceptable residual error rate. The transport protocol contains a retransmission mechanism for erroneous data units based on a retransmission timer. Under these conditions, it is possible to reduce the upper window limit with a regular acknowledgement, since the data units lost during this operation will be detected by the timer and retransmitted. This type of operation is represented in Figure 3.34 where the upper window limit is reduced by acknowledgement AK4, CDT = 0; this causes blocking of the sender and the loss of TPDU DT4. On expiry of the guard time $T1$ of DT4, the sender detects the loss of DT4 and decides to retransmit this data unit. Retransmission clearly can be performed only after unblocking, which is initiated here by receipt of an acknowledgement AK4, CDT = 1, which offers a credit of one unit.

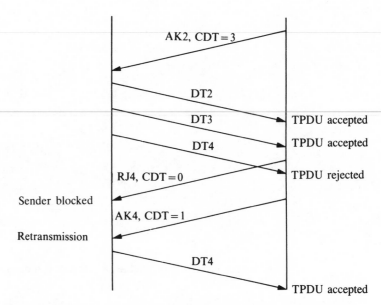

Figure 3.33 Reduction of the upper window limit by a reject TPDU

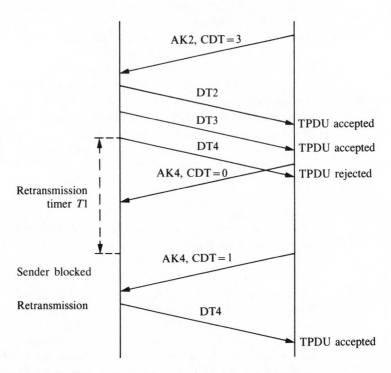

Figure 3.34 Example of reduction of the upper window limit with a data transfer procedure with retransmission

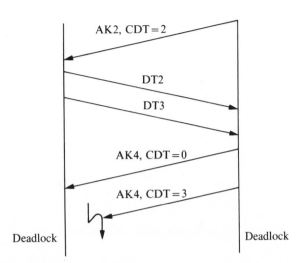

Figure 3.35 Example of deadlock caused by loss of an acknowledgement

Operation of the flow control mechanism for unreliable networks can be compromised by the loss of an acknowledgement. Hence, with the example of Figure 3.35, transfer of data units DT2 and DT3 is correctly acknowledged by AK4, CDT = 0, which blocks all new transmission until a new acknowledgement increases the upper window limit. After some time, the destination again has free receive buffers which permit it to offer the sender a transmission credit of 3 data units by AK4, CDT = 3. If the latter acknowledgement is lost in transit, the sender remains blocked. As the destination has sent a transmission credit, it is justified in thinking that not receiving TPDUs is a normal state of affairs arising from the fact that the sender has nothing to send. There is, therefore, a *deadlock* which is due to a difference of perception of the window control information between the sender and the destination.

The deadlock which is liable to occur while increasing the upper window limit can be eliminated by means of a *window timer* which is started when sending an acknowledgement increases the upper window limit (Figure 3.36). This timer is set to a guard time *TW* which must be greater than the sum of the transfer time of an acknowledgement and the transfer time of a data unit in the other direction. If no data unit arrives at the destination before expiry of the guard time, the destination must conclude that an acknowledgement has been lost and retransmit it.

The window timer must necessarily be set to a high guard time to take account of the fact that the sender may not have any waiting TPDU at the time of opening the credit. If the guard time *TW* is too short, there is a chance that the destination will retransmit acknowledgements at a high rate for a long time, which needlessly uses network resources. On the other hand, too high a guard time leads to needless lengthening of the sender blocking time when this has TPDUs awaiting transmission. It can be seen that the window timer eliminates the deadlock only at the price of a loss of efficiency of the protocol.

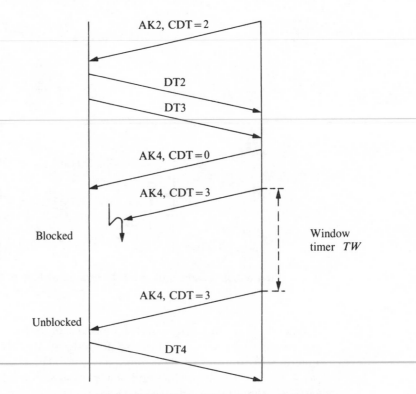

Figure 3.36 Window timer mechanism

Operation of the protocol can be notably improved by introducing a mechanism for acknowledging receipt acknowledgements which increase the upper window limit. With this technique, the destination sets a regular retransmission timer when it sends an acknowledgement which increases the upper window limit. The guard time of the timer is set to a value $T1$ which is just greater than the round trip transit time of an acknowledgement on the network. When it receives an acknowledgement which increases its upper window limit, the sender of the DT TPDUs must respond by returning an acknowledgement which contains, in its supplementary parameter field, the values of $T(R)$ and CDT carried by the acknowledgement which it has just received (Figure 3.37). This 'acknowledgement of an acknowledgement' permits the destination to ensure that the sender operates with the most recent flow control parameters and causes it to retransmit the acknowledgement if it has not received confirmation at the end of a delay $T1$. The delay in updating the window in the case of loss of an acknowledgement is thus reduced to a minimum.

The preceding protocol can fail when acknowledgements are received out of sequence. Consider the example of Figure 3.38 where data units DT2 and DT3 are acknowledged by AK4, CDT = 3. Following a sequencing failure on the network, AK4, CDT = 3, is received before AK4, CDT = 0, so that the sender immediately returns a **_flow control confirmation_** by AK5, CDT = 7, (AK4, CDT = 3). The destination wrongly reads this confirmation as an indication that

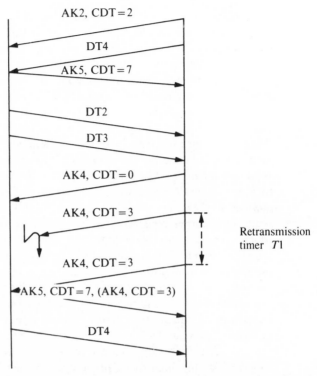

Figure 3.37 Example of flow control procedure using a flow control confirmation parameter

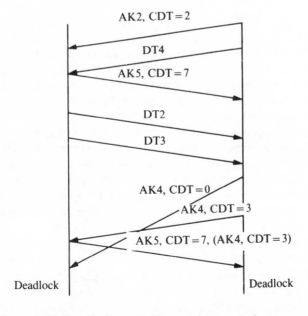

Figure 3.38 Deadlock caused by receiving out-of-sequence acknowledgements

the sender has a credit of 3 units, when this actually has zero credit since the last received acknowledgement contained the value CDT = 0. This is again a deadlock. A first solution to prevent deadlock is to abandon the flow control confirmation technique and again use the approach based on periodic retransmission of acknowledgements monitored by a window timer. In this case, the destination will always end up by transmitting to the sender an acknowledgement containing the flow control parameters which will re-establish synchronism.

Another solution to the deadlock problem posed by out-of-sequence transmission of acknowledgements is to assign transmission sequence numbers to the acknowledgements. With this method, the acknowledgement contains, in addition to the number $T(R)$ of the next expected TPDU, a *transmission sub-sequence number SN(S)* which should be incremented by one each time the transport entity sends an acknowledgement. As it is sufficient to distinguish an acknowledgement which increases the upper limit of the window from the previous one, the sub-sequence number is incremented by one only when the acknowledgement increases the upper window limit. The sub-sequence numbers are reset to zero each time the receive sequence number $T(R)$ is changed. On the transmission side, examination of the sub-sequence number of the received acknowledgements enables those which arrive late following a sequence failure to be discarded, and hence the flow control parameters to be maintained at their correct values. It can be seen that sub-sequence numbering can complement the flow control confirmation mechanism by preventing the deadlocks which this could involve.

3.2.5 Data transfer mechanisms

A number of the data transfer procedures have been described in connection with flow control. These procedures cover sequencing, error detection and correction and flow control. The corresponding mechanisms are closely linked, although the credit allocation method introduces decoupling between flow control and error detection and correction. The overlapping of the data transfer protocol mechanisms means that in practice the acknowledgements serve simultaneously for sequencing, flow control and error detection and correction; this can be a source of problems as will be seen in the examples in this section.

Recall firstly that the DT data units can be duplicated when the acknowledgement is lost in an error correction by retransmission procedure (Figure 3.30). At the destination, the duplicated DT TPDUs can easily be eliminated, since their sequence number is identical to that of a data unit received previously. The duplicated DT TPDUs must clearly be discarded but it is necessary for them to be acknowledged so that the sender frees the corresponding transmission buffer.

The dual role played by the acknowledgements in connection with error correction and flow control can be the source of a protocol error when a DT TPDU is duplicated. Consider the example of Figure 3.39 where the TPDU DT0 is correctly received by the destination transport entity B, but the corresponding acknowledgement AK1 is lost. The source transport entity A retransmits DT0, but as this data unit wanders in the network, it reaches the destination only after

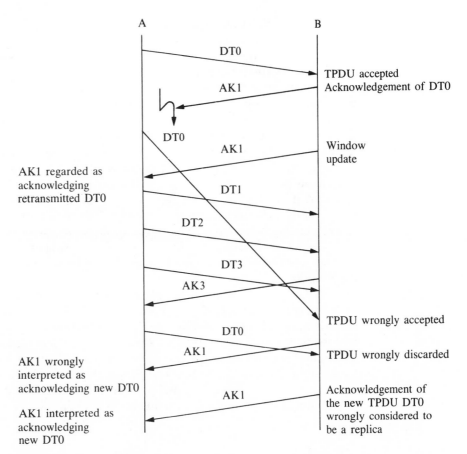

Figure 3.39 Protocol error caused by the delay of a duplicated data unit. Sequence numbering modulo 4

a very long delay. Meanwhile, destination B is justified in believing that its first acknowledgement AK1 has been correctly received and is caused, after a certain time, to retransmit a new acknowledgement AK1 to update the window parameters. This new acknowledgement AK1 is wrongly read by sender A as an acknowledgement of data unit DT0 which it has just retransmitted so that it considers the situation to be normal and discards the replica of DT0 which it was saving. After these events, sender A successively sends data units DT1, DT2, DT3 then TPDU DT0 since sequence numbering is performed modulo 4. If the duplicated and delayed TPDU DT0 arrives at the destination between TPDU DT3 and the new TPDU DT0, it will be wrongly accepted since it arrives in sequence. When the new TPDU DT0 arrives at the destination, the latter wrongly considers it to be a replica of the first which causes it to discard it and return an acknowledgement. The sequence of events thus leads to a protocol error in which an old data unit is substituted for a more recent one which carries the same send number.

There are several possible solutions to the problem posed by the rogue data units. A first prevention is to choose a sequence numbering modulus sufficiently high so that the arrival in sequence of a delayed data unit corresponds to a delay greater than the maximum network transfer time. This technique is used effectively and it will be seen that the ISO transport protocol provides modulo 2^7 sequence numbering with optional modulo 2^{31} numbering. In the latter case, an error caused by cyclic numbering is clearly very unlikely but it is not quite impossible since a packet can always have been delayed for a considerable time in one of the queues at a network node. Use of high modulo numbering must, if possible, be combined with a limitation of packet lifetime where the network practises euthanasia by eliminating very old packets.

So far, it has been assumed that error detection and correction is performed using positive acknowledgements transmitted by the destination transport entity to the sending transport entity in the form of AK data units. If there is no multiplexing, it is possible to move the acknowledgement mechanism to the network level. With this variant of *implicit acknowledgement*, which is called **confirmation of receipt**, the DT TPDUs to be sent are submitted to the network by an N-DATA request primitive (NDTreq) which contains a *confirmation request* as a parameter (Figure 3.40). On the destination side, the arrival of the corresponding packet causes an N-DATA indication primitive (NDTind) to be sent to the transport layer; this also contains a *confirmation request* as a parameter. The transport entity must then transmit an acknowledgement by the network layer by means of an N-DATA ACKNOWLEDGE request primitive (NDTAKreq). The arrival of the corresponding acknowledgement packet causes an N-DATA ACKNOWLEDGE indication primitive (NDTAKind) to be sent to the sending

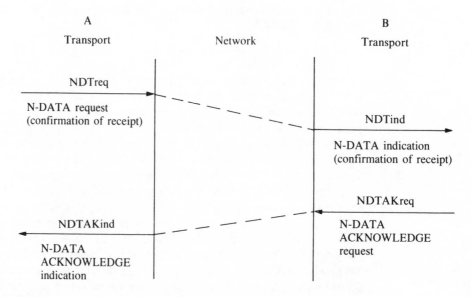

Figure 3.40 Implicit acknowledgement by confirmation request at the network level

transport entity; this permits acknowledgement of the TPDU and clearing of the corresponding transmission buffer.

Transport protocols do not usually provide the facility of piggybacking the acknowledgement of a DT TPDU in a DT TPDU sent in the other direction. Such an approach would complicate the data transfer procedures even more without notably improving the efficiency of network utilization. It has been seen that it is possible to combine several TPDUs into a single packet at the network/transport interface. Under these conditions, sending of an acknowledgement can be combined with that of a user data unit by *concatenation of a DT TPDU and an AK TPDU into the same NSDU*, which comes close to piggybacking the acknowledgement in a DT TPDU.

The type of fault likely to affect a data unit during its transfer varies with the type of network and data links. If the data links correct errors perfectly, faults are due essentially to breakdowns on the data circuits or faults in the network switching-multiplexers. These failures are, in principle, signalled by the network to the transport layer and failures which are not signalled lead directly to the loss of data units. In this case, it is superfluous to protect the data units by an error-detecting code. It is, in contrast, necessary to provide error detection with a supervisory field in the data units when the unsignalled error rate on the network is unacceptable. It will be seen that the ISO transport protocol offers this facility as an option.

It can happen that a transport connection can be closed at one of the transport entities although it remains open for the opposite entity. This event can occur, for example, following the loss of disconnect TPDUs or as a consequence of a breakdown of the equipment executing one of the transport entities. To take care of this problem, operation of the connection is monitored by an **inactivity timer** which is reset on each receipt of a valid TPDU. The timer is set to a guard time TI and the transport entity starts a connection clearing procedure if no TPDU has been received before expiry of the guard time. In order to prevent untimely clearing when no data is sent, the inactivity guard time TI must be greater than the guard time TW of the window timer. In the absence of data transfer, this ensures transmission of a sufficient number of AK TPDUs on a sound connection to keep the connection open.

3.2.6 Transport connection clearing mechanisms

In normal operation, a transport connection can be cleared abruptly (*abrupt termination*) on the initiative of any one of the users of the transport layer who sends it a T-DISCONNECT request primitive (TDISreq). On arrival of this primitive, the transport entity discards the TPDUs awaiting transmission and sends a *disconnect request* data unit DR (Figure 3.41). From this time, the transport entity ignores all received TPDUs except for a DR or DC TPDU. At the distant transport entity, the DR TPDU causes clearing of the connection with return transmission of a *disconnect confirm* DC TPDU and sends to the local user a T-DISCONNECT indication primitive (TDISind) which contains an indication of the reason for clearing as a parameter. The process terminates with

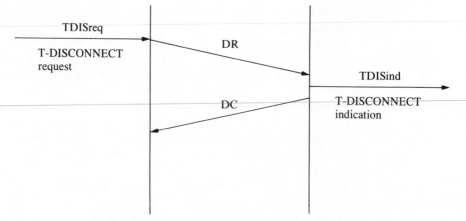

Figure 3.41 Normal transport connection clearing

the arrival at the sender of a DC data unit which confirms clearing of the connection and causes a *freezing of the reference* corresponding to the terminated connection, if this option is used.

The clearing procedure described above is that which is used with the ISO transport protocol. It can lead to a loss of data at the time of clearing if users do not take appropriate action before requesting clearing. Some protocols, such as that proposed by the *National Bureau of Standards* (*NBS*) which is a variant of the ISO transport protocol, propose an optional *graceful close* procedure which prevents data loss at the transport level during connection clearing.

The NBS graceful close mechanism is illustrated in Figure 3.42 where the T-CLOSE request primitive causes sending of a GR (*graceful close request*) data unit which is sent only after all the DT TPDUs waiting for transmission are sent

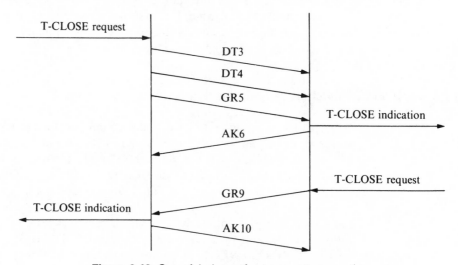

Figure 3.42 Graceful close of a transport connection

and is numbered in sequence with the latter. The transport entity no longer accepts new data transfer requests but it leaves the transport connection established in order to receive the TPDUs still in transit. The distant entity checks, by means of the sequence number of GR, that it has indeed received all the TPDUs in transit on the connection. If this is the case, it returns an acknowledgement to the sender and informs its local user of the closure with a T-CLOSE indication primitive. The arrival of the acknowledgement at the sender of the GR TPDU effectively closes the half connection in the direction where GR has been transmitted. The other half connection is closed in an ordered manner according to an identical principle.

3.2.7 Recovery following reset of the network connection

When the network is subject to failures, it can happen that it initiates reset of an established network connection following, for example, congestion or an irrecoverable sequence error. This reset is signalled to the two ends of the network connection by an N-RESET indication primitive (NRSTind) to which each of the two transport entities must reply with an N-RESET response (NRSTrsp) (Figure 3.43). This resetting corresponds to a resynchronization where none of the data units sent before the sender reset are received after reset of the destination. Conversely, any data unit transmitted after the sender reset cannot be received before reset of the destination. Finally, some data units may be lost during reset.

Reset of a network connection must clearly cause resynchronization of the transport connections or, by default, clearing of the transport connections if resynchronization proves to be impossible. The resynchronization procedure used

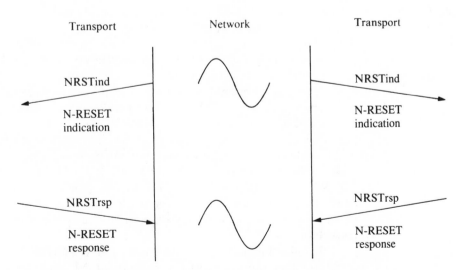

Figure 3.43 Service primitives exchanged at the time of network reset on the initiative of the network service provider

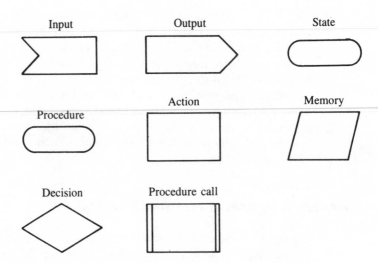

Figure 3.44 Graphic symbols of the SDL language

in ISO transport protocol 8072–8073 classes 1 and 3 will be briefly described here by using the symbols of the *specification and description language (SDL)* which are shown in Figure 3.44.

As reset is signalled to the two ends of the network connection, the first problem which arises is to determine which of the two transport entities must take the initiative in resynchronizing the transport connection. The convention adopted is that resynchronization must be provided by the entity which initially established the transport connection. On the other hand, the time devoted to resynchronization must be limited since successful conclusion of the operation becomes more and more unlikely as time passes and it is wasteful to operate a transport connection which is not providing a useful service for too long. Resynchronization is, therefore, monitored by the initiator of the transport connection using a *time to try reassignment/resynchronization timer* whose guard time *TTR* must not exceed about 2 minutes. Similarly, resynchronization at the called entity is monitored by a *time to wait for reassignment/resynchronization timer* whose guard time *TWR* is set to a value greater than *TTR*.

In the steady state, the transport connection is in the *open* state. On the arrival of an N-RESET indication primitive, NRSTind, the transport entity checks that it did establish the transport connection supported by this network connection. If this is the case, the transport entity sets the *TTR* timer, sends a reject data unit RJ and returns an N-RESET response primitive (NRSTrsp) to the local network layer (Figure 3.45). After these operations, the transport connection remains in the *open* state while waiting to receive an RJ TPDU which will achieve resynchronization.

At the other end of the network connection, the NRSTind primitive causes the corresponding transport entity to return an NRSTrsp primitive to the local network entity while leaving the transport connection in the *open* state. The transport entity then waits for the arrival of an RJ data unit sent by the entity

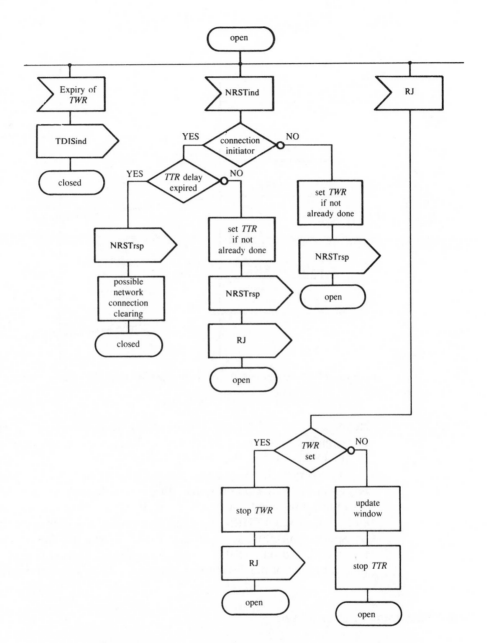

Figure 3.45 Diagram of operation as a result of network reset

which initially established the transport connection. If this event occurs before
expiry of the *TWR* timer, the called entity stops the *TWR* timer and replies with
an RJ TPDU to complete synchronization. After these operations, the transport
connection remains in the *open* state.

At the transport entity which initially established the connection, receipt of the
expected RJ TPDU completes resynchronization, which terminates with the

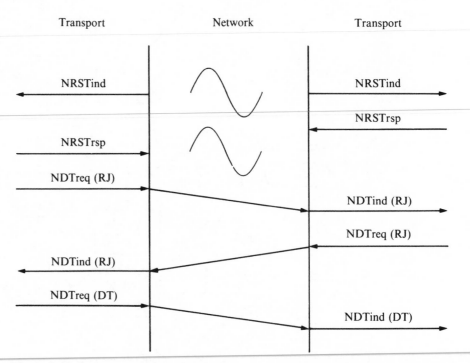

Figure 3.46 Sequence of resynchronization operations as a result of network reset

stopping of the *TTR* timer (Figure 3.46). Data transfers can then restart from the sequence numbers indicated by the RJ data units; this permits retransmission of the TPDUs lost at the time of reset.

It can very well happen that the *TTR* timer has already been started before receipt of the NRSTind primitive, for example following a network disconnection. This timer thus need not be restarted. If the guard time is expired, the transport entity must send an NRSTrsp primitive to the local network entity and clear the transport connection. It could also cause clearing of the network connection if this is allocated only to the cleared transport connection (Figure 3.45). Expiry of the guard time of the *TWR* timer also causes clearing of the transport connection.

3.2.8 Reassignment after a fault

The network can be subject to serious faults which cause unexpected clearing of one or more network connections with signalling of the fault to the two ends of the connection by N-DISCONNECT indication primitives (NDISind) (Figure 3.47). In such a case, several strategies are possible. The most direct method is to clear the transport connections using the network connection whose clearing is signalled. This approach has the merit of being simple, but it has the serious disadvantage of causing the loss of the data which was in transit in the transport layer; this is contrary to the rules of the OSI model since the transport layer

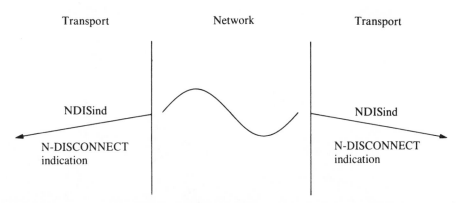

Figure 3.47 Service primitives exchanged at the time of network disconnection on the initiative of the network service provider

must provide reliable end-to-end transport. It is, therefore, generally necessary to adopt a policy of systematic attempts to reassign the transport connections affected by the disconnection to other network connections, if it is possible to find sound ones.

It can well happen that two communicating transport entities use several network connections in parallel. If one of these is released following a fault, it is possible to reassign the transport connections to network connections that still operate, using multiplexing and accepting a reduced rate if necessary. If no network connection is available, the transport entities must attempt to establish a new network connection to reassign the orphaned transport connections. After reassignment, it is clearly necessary to start a resynchronization procedure in such a way as to retransmit the data lost at the time of the fault and re-establish the correct sequence.

The principle of the reassignment procedure of ISO transport protocol 8072–8073, classes 1 and 3, will be briefly presented here; it uses the same *TTR* and *TWR* timers as for resynchronization. By convention, the initiative for reassignment operations also reverts to the transport entity which established the connection; for simplicity this will be called the *calling entity*, in contrast to its correspondent which will be called the *called entity*.

Monitoring of reassignment is performed at the calling entity by the *TTR* timer. When clearing of the network connection is signalled by the NDISind primitive, the calling entity sets the *TTR* timer and checks whether there is another available network connection towards the called entity. If this is the case, the calling entity assigns the transport connection to this network connection and starts a resynchronization procedure by sending an RJ TPDU. At the end of these operations, the transport connection remains in the *open* state (Figure 3.48). At the other end of the connection, the called entity sets the *TWR* timer and puts the transport connection into an intermediate *open*-WR state when the NDISind primitive signals clearing of the network connection to it. If an RJ TPDU arrives at the called entity before expiry of the guard time *TWR*, the called entity returns an RJ data unit which terminates the reassignment/resynchronization operations

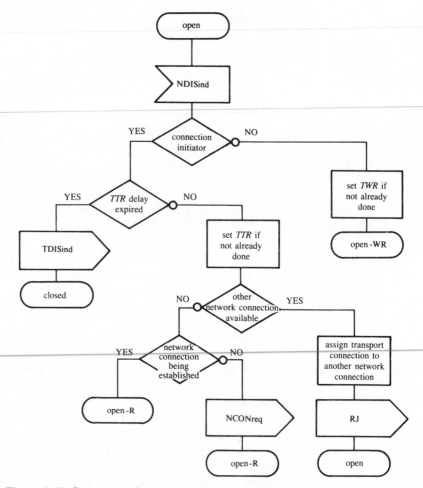

Figure 3.48 Sequence of events caused by a network disconnection indication

after stopping *TWR* (Figure 3.49). When the guard time *TWR* expires before the arrival of an RJ TPDU, the procedure is aborted and the transport connection must be cleared.

When the calling entity does not have another available network connection at the time disconnection is signalled by NDISind, it must establish a new connection. If this is already in course of establishment, the calling entity puts the transport connection into an intermediate *open*-R state. In the other case, the calling entity attempts to obtain a new network connection by sending an N-CONNECT request primitive (NCONreq) to the local network entity (Figure 3.48). Establishment of the new network connection is signalled to the calling entity by an N-CONNECT confirm primitive (NCONcnf) which leads the calling entity to reassign the transport connection to the new network connection and to send an RJ TPDU to resynchronize (Figure 3.50). When the guard timer *TTR* expires before the arrival of the NCONcnf primitive, the calling entity must assume that the network has been unable to establish a new connection and it must clear the

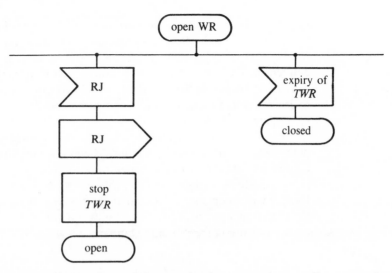

Figure 3.49 Waiting for reassignment by the transport connection responder

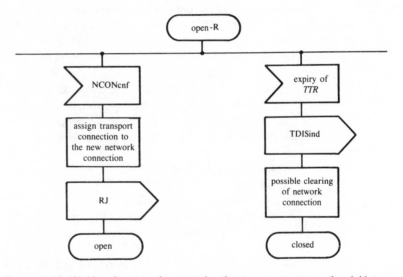

Figure 3.50 Waiting for reassignment by the transport connection initiator

transport connection by signalling this event to the transport layer user with a TDISind primitive.

3.3 CONNECTION MODE TRANSPORT LAYER ISO 8072–8073

3.3.1 Introduction

The transport layer with connection specified by ISO in International Standards ISO 8072 for the *connection-oriented transport service definition* and ISO 8073 for the *connection-oriented transport protocol specification* will be described here [3.14]. These standards are identical to CCITT recommendations X.214 and X.224 respectively [3.21]. They are derived from preliminary work by ECMA which defined the ECMA-72 transport protocol [3.16]. In the United States, the *National Bureau of Standards* has specified an NBS transport protocol which is very similar to ISO protocol classes 2 and 4 but has some differences which will be detailed subsequently [3.15].

The ISO transport protocol, which is very complex, will not be described in detail here; its main procedures have been presented in the previous section. The reader who wishes to go deeper in this area could usefully consult the ISO standard which contains a *state table* for the various protocol classes. The NBS standard is much more detailed than the ISO standard and can be very useful for clarifying some aspects of the ISO protocol whose interpretation is not always clear from the corresponding standard.

3.3.2 Protocol classes

It has been seen that network connections can be classified into three types, A, B and C, according to their quality, where type A corresponds to a reliable network, type B corresponds to a network with unacceptable rate of signalled failures and type C defines an unreliable network likely to have unsignalled failures (Table 3.1).

Transport entities must know the type of available network connections in order to operate procedures which permit them to ensure the required quality of service. To simplify the choice of these procedures, they are grouped into five classes numbered from 0 to 4 and whose functions are as follows:

Class 0: Simple class
Class 1: Basic error recovery class
Class 2: Multiplexing class
Class 3: Error recovery and multiplexing class
Class 4: Error detection and recovery class

Protocol classes are generally associated with network connections as indicated in Table 3.2.

Class 0 corresponds to the simplest possible transport protocol. It was originally developed for the text transmission *Teletext* service which can be considered as an improved version of the *telex* service. Class 0 provides connection establishment and data transfer with segmentation functions (Table 3.3). In this case there are no multiplexing, explicit flow control or error correction functions and transport connection release is performed implicitly by clearing the network connection.

Class 1 was originally developed by the CCITT to operate with a type X.25 network. Because of this, multiplexing was not provided, since X.25 offers a very large number of distinct virtual circuits. In this class, the DT TPDUs are numbered and the protocol operates a DT TPDU retention mechanism until their acknowledgement; this permits resynchronization after reset of the X.25 network

Table 3.2 Use of transport protocol classes as a function of network connection type.

	Type of network connection		
	A	B	C
Acceptable residual error rate	yes	yes	no
Acceptable rate of signalled failures	yes	no	no
Class to be used	0 without multiplexing 2 with multiplexing	1 without multiplexing 3 with multiplexing	4

Table 3.3 Assignment of procedure elements in each class.

	Procedure class				
Protocol mechanism	0	1	2	3	4
Assignment to a network connection	×	×	×	×	×
TPDU transfer	×	×	×	×	×
Segmentation and reassembly	×	×	×	×	×
Concatenation and separation		×	×	×	×
Connection establishment	×	×	×	×	×
Connection refusal	×	×	×	×	×
Normal release Explicit		×	×	×	×
Implicit	×				
Release on error	×		×		
Association of TPDUs with transport connections	×	×	×	×	×
TPDU DT numering		×	×	×	×
Expedited data Normal network		×	×	×	×
Expedited network		×			
Reassignment after failure		×		×	
Retention until TPDU AK		×		×	×
acknowledgement confirmation of receipt			×		
Resynchronization		×		×	
Explicit flow control	with-out	with-out	with or without	with	with
Error detection by checksum (option)					×
Frozen reference		×		×	×
Retransmission on timeout					×
Resequencing					×
Inactivity control					×
Protocol error processing	×	×	×	×	×
Splitting and recombining					×
Multiplexing and demultiplexing			×	×	×

(X.25 RESET). The DT TPDU are saved by the sender until their correct delivery is confirmed by an AK TPDU or, as an option, by implicit confirmation from the network. In the latter case, the D bit, *delivery confirmation*, of the X.25 packets is set to 1 to indicate that data packet acknowledgement originates from the remote network entity. As is expected for a class without multiplexing, flow control is provided not at transport level but at network level in an implicit manner. Class 1 is provided for type B network connections.

Class 2 is an improved version of class 0 which is also designed for type A network connections but, in this case, with a function for multiplexing several transport connections onto one network connection. The TPDU must thus contain a *reference* to distinguish the various transport connections multiplexed on the same network connection, together with a sequence number to permit flow control by a *credit mechanism* (§ 3.2.4). As an option, sequence numbering can remain unused with flow control performed in an implicit manner in the network layer; this clearly has the disadvantage of spurious coupling between the rates of the various transport connections multiplexed on the same network connection. As the network connections are considered to be perfectly reliable in this case, class 2 does not contain error detection and recovery functions. Because of this, reset and release of the connection in class 2 cause disorderly release of the transport connection without end-to-end exchanges between the two distant transport entities.

Class 3 is intended for use with type B network connections which are subjected to signalled resets and breaks. In simple terms, this class combines the functions of classes 1 and 2 in the sense that it permits both multiplexing and recovery after a signalled failure.

Class 4 is intended to operate with type C network connections whose residual error rate is unacceptable. This class is the most advanced of all and contains all the functions of the other classes with the exception of resynchronization and reassignment after a failure since these signalled failures are processed in the same way as unsignalled failures. Class 4 can be used with a connectionless network operating with a datagram type of service; consequently, its use suits local networks where exchanges are often performed without connection. In addition to the preceding functions, there are also mechanisms for error correction by retransmission, inactivity detection and resequencing. The basic procedures are intended for detection and correction of lost or duplicated TPDUs. When the network is also liable to produce transmission errors, these can, optionally, be detected and corrected by using an error-detecting code with a 16-bit checksum in the DT TPDUs.

3.3.3 Transport and network service primitives

The transport layer provides services for the session layer. Conceptually, these services correspond to the exchange of transport service primitives which are exchanged locally at the interface between the transport layer and the session layer. These primitives were described in the previous section in connection with the elementary transport protocol mechanisms. A complete list is given in Table

Table 3.4 Transport service primitives.

Primitive	Symbol	Parameters
T-CONNECT request	TCON req	(called address, calling address, expedited data option, quality of service, user data)
T-CONNECT indication	TCON ind	"
T-CONNECT response	TCON rsp	(quality of service, responding address, expedited data option, user data)
T-CONNECT confirm	TCON cnf	"
T-DATA request	TDT req	(user data)
T-DATA indication	TDT ind	"
T-EXPEDITED DATA request	TEX req	"
T-EXPEDITED DATA indication	TEX ind	"
T-DISCONNECT request	TDIS req	(user data)
T-DISCONNECT indication	TDIS ind	(disconnect reason, user data)

3.4. It can be seen that the only new primitives are those which concern expedited data transfer with a T-EXPEDITED DATA request (TEXreq) which causes a T-EXPEDITED DATA indication (TEXind) primitive to be sent to the distant user (Figure 3.51). These primitives play the same role for expedited data as normal data transfer primitives.

Allowed sequences of transport service primitives are shown in Figure 3.52 with four possible states at the transport/session interface for the transport connection.

The parameters of the various primitives which are shown in Table 3.4 are only those which are independent of the particular realization of the transport/session interface. They must also include information which is exchanged in a manner depending on the particular interface implementation. Hence, for example, the data transfer primitives must indicate the *reference* of the transport connection.

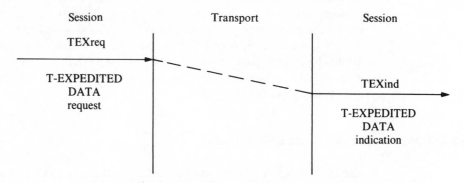

Figure 3.51 Expedited data transfer primitives

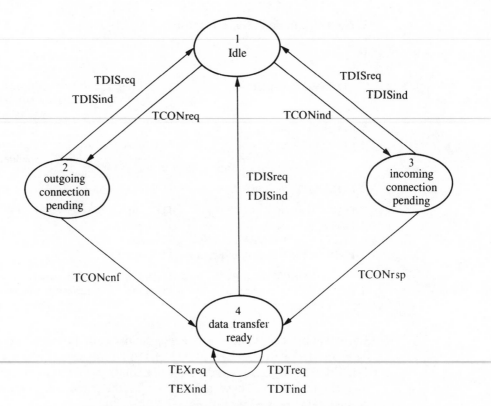

Figure 3.52 Allowed sequences of transport service primitives

Similarly, the local flow control mechanism at the transport/session interface must be specified in the design phase.

The transport layer uses the network layer services with exchanges which are conceptually defined by the network service primitives recalled in Table 3.5. These primitives play the same role at network level as the corresponding transport service primitives as far as connection establishment, normal or expedited data transfer and disconnection are concerned. Additional primitives are needed for network reset on the initiative of the transport entity or the network itself; the use of these has been partly described in § 3.2.7 devoted to recovery after reset. The N-DATA ACKNOWLEDGE request and indication primitives serve to acknowledge data packets at network level; in certain cases this permits transport data units to be acknowledged in an implicit manner by confirmation of receipt at network level (Figure 3.40).

3.3.4 Transport protocol data unit formats

The ISO transport protocol uses ten different types of TPDU which can themselves be divided into subtypes according to the class of protocol (Table 3.6). The

Table 3.5 Network service primitives.

Primitives	Symbol
N-CONNECT request	NCON req
N-CONNECT indication	NCON ind
N-CONNECT response	NCON rsp
N-CONNECT confirm	NCONcnf
N-DATA request	NDT req
N-DATA indication	NDT ind
N-DATA ACKNOWLEDGE request	NDTAK req
N-DATA ACKNOWLEDGE indication	NDTAK ind
N-EXPEDITED DATA request	NEX req
N-EXPEDITED DATA indication	NEX ind
N-RESET request	NRST req
N-RESET indication	NRST ind
N-RESET response	NRST rsp
N-RESET confirm	NRST cnf
N-DISCONNECT request	NDIS req
N-DISCONNECT indication	NDIS ind

functions covered by the various TPDU have already been seen in the previous section with the exception of those which correspond to the ED, EA (expedited data) and ER (protocol error report) TPDUs; their use will be described subsequently in this section.

Transport protocol data units have a variable format with the general structure indicated in Figure 3.53. Each TPDU starts with a header whose size depends

Table 3.6 TPDU types. O = option.

Symbol	Name	Valid in classes				
		0	1	2	3	4
CR	Connection request	×	×	×	×	×
CC	Connection confirm	×	×	×	×	×
DR	Disconnect request	×	×	×	×	×
DC	Disconnect confirm		×	×	×	×
AK	Data acknowledgement		O	O	×	×
EA	Expedited data acknowledgement		×	O	×	×
DT	Data	×	×	×	×	×
ED	Expedited data		×	O	×	×
ER	TPDU error	×	×	×	×	×
RJ	Reject		×		×	

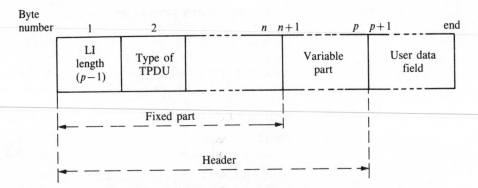

Figure 3.53 General structure of transport protocol data units

on the type of data unit and the options selected at the time of establishment. The header may be followed by a user data field. The header contains a fixed part whose length depends only on the type or subtype of TPDU and a variable part which contains the various additional parameters selected for the transport connection.

The fixed part starts with a byte LI which indicates the length of the header, expressed as a number of bytes, not including byte LI itself. The LI field is followed by a second byte whose first four bits define the TPDU type. The last four bits of the second byte can indicate the credit CDT for CR, CC, AK and RJ TPDUs and are set to zero in other cases. The rest of the fixed part of the header depends on the type of data unit and will be presented later in this section.

The variable part of the header contains parameters which are negotiated at the time of connection establishment. These parameters are coded as indicated in Figure 3.54 with a first byte which defines the parameter using a pre-established code. This first subfield is followed by a second byte which specifies the length of the parameter expressed as a number of bytes. The actual parameter value is indicated by a last subfield of variable length.

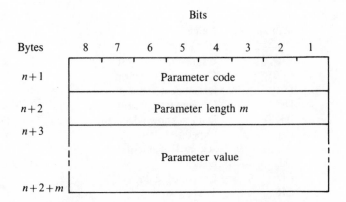

Figure 3.54 Coding of parameters of the variable part of the header

The TPDU must contain a whole number of bytes and the header can be followed by a data field whose size is variable. The size of the data field is not indicated in the header, since the TPDU must be routed as a whole by the network layer; hence the data field is defined without ambiguity as the part which exists after the header.

In the case of unreliable network connections, that is for class 4 alone, all TPDUs can contain a **checksum** in the variable part of the header; this is a 16-bit parameter which is used for error detection and is computed on the whole TPDU according to an algorithm which will be described in § 3.3.9.

The *connection request* data unit CR has the general format shown in Figure 3.55. The last four bits of the second byte contain the initial credit allocation CDT, except for classes 0 and 1 for which the credit mechanism cannot be used (no multiplexing). The first two bytes of the CR TPDU are followed by two 16-bit fields which contain the *destination reference* and the *source reference* respectively. The *source reference* identifies the transport connection chosen by the transport entity which sent the CR TPDU. The *destination reference* is that which will be chosen by the distant transport entity following receipt of the CR TPDU. It is thus still unknown and this causes the sender to set a zero value in this field. The *references* are followed by a byte of which the first subfield indicates the *sender's preferred transport class* for the requested connection. The second subfield of four bits indicates the *options* requested by the sender which can be the use of normal (modulo 2^7) or extended (modulo 2^{31}) numbering in classes 2, 3 and 4 together with the use or not of explicit flow control in class 2. The variable part of the header can contain the parameters indicated in Table 3.7.

The *connection confirm* data unit CC has the same format as the CR TPDU since it plays a similar role to the latter in establishing the connection. The destination *reference* is, in this case, identical to the source *reference* of the CR TPDU and it thus identifies the transport connection of the initiator of the transport connection. The source *reference* of the CC TPDU indicates the transport connection chosen by the transport entity sending the CC TPDU.

Except for class 0, CR and CC TPDUs can contain a user data field whose length is limited to 32 bytes; this can be used to transfer information associated with initialization, for example a password.

The *disconnect request* DR TPDU contains the source and destination *references* together with a one-byte field which indicates the *cause* of the disconnection request; this can, for example, be a protocol error or an invalid header length. The DR data unit can contain a checksum in the variable part of the header together with a parameter containing additional information relating to disconnection. The header can be followed by a user data field whose length must not exceed 64 bytes and whose error free transfer is not guaranteed.

The *disconnect confirm* data unit DC is not used in class 0, since this provides transport connection clearing by closing the network connection without an exchange between the two transport entities. The DC TPDU contains the source and destination *references* and possibly a checksum in its variable part.

The *data transfer* DT TPDU is available in three versions which are chosen according to the protocol class and the selected options (Figure 3.55). With the simplest version, which can be used only in classes 0 and 1, the data unit DT

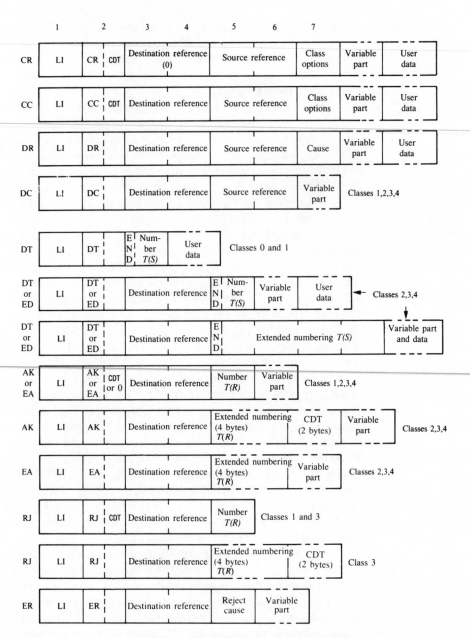

Figure 3.55 Format of transport protocol data units

contains only one additional byte in its header after the length indication LI and the TPDU type identification. This byte includes the 7 bits of the send number $T(S)$ (modulo 2^7) together with an END bit which indicates, when at 1, that the data unit is the last of a message. This DT TPDU does not contain a variable part and clearly finishes with the user data field. Notice that this data unit does not contain a *reference*; this is logical since classes 0 and 1 do not support

Table 3.7 CR TPDU parameters.

Parameter	Significance
Transport service access point identifier (TSAP-ID)	Necessary when the network address is insufficient to determine the transport address (Figure 3.7)
TPDU size	Maximum proposed TPDU size (default value = 128 bytes)
Version number (not used in class 0)	Version of the transport protocol used for the connection
Security parameter (not used in class 0)	User defined
Checksum (used only in class 4)	Error check word; must always be included in class 4 even if the checksum option is not requested
Additional option selection (not used in class 0)	Expedited network option or not in class 1; AK option or confirmation of receipt in class 1; checksum used or not (class 4); expedited data transport service or not
Alternative protocol classes (not used in class 0)	Classes which can be selected by the remote entity if the preferred class cannot be chosen
Acknowledge time (used only in class 4)	Maximum acknowledge time A_L proposed to the remote entity
Throughput (not used in class 0)	Maximum and mean throughput, caller to called direction and vice versa; target value, minimum acceptable value (byte/s)
Residual error rate (not used in class 0)	Target value; minimum acceptable; size of reference TSDU
Priority (not used in class 0)	0 = highest priority
Transit delay (not used in class 0)	Caller–called direction and vice versa (target value, maximum acceptable); time in milliseconds for a 128 byte TSDU
Reassignment time (used only in classes 1 and 3)	Binary value of TTR expressed in seconds

multiplexing and there is, therefore, no ambiguity in the transport connection used. On the other hand, sequence numbering is justified only in class 1, for recovery after a signalled failure; it is of no use in class 0 which is provided for reliable network connections. The sequence field is thus unused in the case of class 0 operation.

The second version of the DT TPDU is intended for multiplexed transport connections in classes 2, 3 and 4. The data unit includes the same fields as previously plus a destination *reference* field which identifies the transport connection used. With a class 4 protocol, the variable part of the header can consist of a checksum intended for error detection.

For network connections which introduce a substantial delay, it is possible to use a third version of the DT TPDU which permits extended sequence numbering modulo 2^{31} by means of a subfield $T(S)$ of 31 bits.

The *expedited data* ED TPDUs have the same structure as the normal DT data units applicable to classes 2, 3 and 4 but in this case the format with normal sequencing can be used in class 1. Sequence numbering is not used in class 2 and for the other classes it must be performed in a manner which is distinct from the sequence numbering of normal data. Finally, the maximum length of the user data field is limited to 16 bytes and the data units can contain a checksum in class 4.

Data acknowledgements AK are available in two versions, with normal or extended sequence numbering, and with the same general structure as the DT data units which they can acknowledge. In order to provide flow control by a credit mechanism, the AK TPDUs contain a credit CDT, which is located in a subfield of 4 bits of the second byte in the case of normal numbering and in a field of 2 bytes with extended sequence numbering. In class 4, the variable part of the header can contain a checksum together with a sub-sequence numbering parameter which is used to process the AK TPDUs in sequence. Finally, a *flow control confirmation parameter* can be used to acknowledge the received AK TPDUs. This parameter contains the lower window edge together with the sub-sequence number and the credit of the opposite entity received in the AK TPDU from the latter.

The *expedited data acknowledgement* EA TPDUs have the same structure as normal data acknowledgements with the sole difference that the variable part can contain only the checksum parameter in class 4 and the data units do not contain credit in this case, since this facility is not used with expedited data.

The *reject* RJ TPDUs contain the destination *reference*, the receive number $T(R)$ and the credit value CDT. In this case there is no variable part.

The TPDU *error* ER is used to report a protocol error to the opposite entity. Apart from the destination *reference*, it contains the cause of rejection which can be an invalid parameter code, an invalid TPDU type or an invalid parameter value. The header can also contain a variable part with a checksum in class 4 and possibly a parameter which contains the header of the rejected TPDU.

3.3.5 Transport connection establishment procedure

The principle of establishing a transport connection has been presented in § 3.2.2. Attention here will be restricted to a brief indication of the approach adopted for negotiations with the ISO transport layer.

The connection is established on the initiative of one of the users with a TCONreq primitive which contains the required quality of service as a parameter (Figure 3.18–3.19). The calling transport entity establishes a network connection with the called entity if none is already established and sends a tranport connection request data unit CR which contains the required choices in the fixed and variable parts of the header, for example the protocol class, the maximum TPDU size and the choice of checksum (Table 3.7).

In order to ensure the success of the negotiations, they must observe precise rules. The policy adopted here is to submit the preferred choices by the CR TPDU. The called entity can accept these choices or make counter-proposals taken from the alternative options indicated in the CR TPDU or which approach a value agreed in advance. These counter-proposals are sent by a CC TPDU to the calling entity which must accept them. With the ISO transport layer, the negotiations relate to the following points:

(a) Protocol class
(b) TPDU size
(c) Normal or extended format
(d) Choice of checksum
(e) Quality of service parameters (throughput, transit delay, priority, residual error rate)
(f) No explicit flow control in class 2
(g) Use of the *network confirmation of receipt* and *expedited network* options in class 1.
(h) Utilization or otherwise of the expedited data transport service

The calling entity expresses its wishes concerning the protocol class applicable to the connection by proposing a preferred class by the CR TPDU and possibly alternative fallback classes in the case where the preferred class cannot be adopted by the called entity. The called entity must respond by its CC TPDU with a class chosen from the proposed classes or with an implicit negotiation; according to this, class 2 is a valid response to a class 3 or 4 proposal and class 0 is a valid response to a class 1 proposal.

In connection with the maximum TPDU size, the protocol provides a maximum default size of 128 bytes. The remote entity must respond to a proposal of maximum size contained in the CR TPDU with a counter-proposal which approaches the default value and hence is between 128 and the proposed value.

For the other parameters, the possible valid responses to the proposals of the calling entity are defined by the standard. In this way, for example, the remote entity cannot oppose the proposal to use the checksum in class 4. In contrast, it can impose the use of a checksum in class 4 even if the sending entity made the opposite proposal. Similarly, the remote entity must accept the standard format if it is proposed by the sending entity and it can decide on the use or otherwise of the extended format only if it has been proposed by the sending entity.

3.3.6 Normal data transfer procedures

Data transfer mechanisms have been examined in § 3.2.3 to § 3.2.5 and will not be repeated here except to specify the procedures applied with the different protocol classes.

The user data is transferred by one or more DT TPDU following a TDTreq request primitive and delivered to the destination by a TDTind indication primitive (Figure 3.26).

In the case of a reliable network connection, that is of type A, the transport protocol is normally of *class 0* without multiplexing. Flow control is then provided by the network, and as the transport connection is unique, the DT TPDUs are of the reduced format type without a connection *reference* (Figure 3.55). On the other hand, the DT TPDUs are not acknowledged so their sequence number $T(S)$ is not used. The class 0 protocol provides segmentation and reassembly in order to permit the use of networks where the packet size is limited.

Class 1 is intended for operation without multiplexing on a network which can be subject to signalled failures. As the transport connection is unique, the DT TPDUs are again of reduced format without *reference*. However, the sequence number $T(S)$ is used here since the sending entity must operate a TPDU *retention procedure* in order to ensure recovery in the case of a signalled failure. Copies of the DT TPDUs are retained by the sender until a positive acknowledgement is received; this can take the form of a *confirmation of receipt* transmitted by the network or, optionally, an AK TPDU acknowledgement. As flow control is again provided by the network, acknowledgements serve only for recovery so the credit field is unused. Recovery in case of reset or release of the network connection is performed by the resynchronization and reassignment procedures which are based on the reject data unit RJ and whose principle has been presented in § 3.2.7 and § 3.2.8.

With a reliable network connection, multiplexing of transport connections imposes the use of a *class 2* protocol. In this case, each DT TPDU must contain a *destination reference* which identifies the connection with which it is associated at the destination entity. There is no retention procedure here, since the connection is reliable. Flow control can again be provided by the network, as in class 0, provided a possible reduction in maximum throughput is accepted. There are no acknowledgement AK TPDUs and sequence numbering is not used. With the *explicit flow control option*, the DT TPDUs must be numbered and flow control is performed by way of AK TPDUs which carry a credit CDT (§ 3.2.4). These acknowledgements serve only for flow control.

The *class 3* protocol is intended for multiplexed operation of transport connections on network connections which are subject to signalled failures. This protocol uses essentially the same procedures of retention, acknowledgement and resynchronization as for class 1. However, the use of multiplexing requires the transport connection to be identified by a *reference* in the TPDU and flow control must be performed explicitly by AK TPDUs which carry a CDT credit and which are also used for acknowledgement of the TPDU replicas retained by the sender.

The *class 4* transport protocol is intended for multiplexed operation of transport connections which have an unacceptable error rate. They combine the previous procedures, with the exception of the resynchronization and reassignment procedures which are restricted to classes 1 and 3, and add a number of additional procedures which enable unsignalled network failures to be processed. In this case, DT TPDUs carry the destination *reference* and are numbered. Flow control is performed individually on each connection by a credit mechanism. The AK TPDUs serve the purposes of sequencing, acknowledgement and flow control. In order to ensure recovery following loss of or error in a DT TPDU, the latter are retransmitted after the expiry of a guard time. Also, to protect against loss or out-of-sequence receipt of acknowledgements, these can contain a sub-sequence number parameter and a flow control confirmation parameter (§ 3.2.4). The DT TPDU can, optionally, be protected against transmission errors by an error detection code which appears as a 16-bit checksum in the data unit. Finally, the class 4 transport protocol must include inactivity detection and window timing procedures.

3.3.7 Expedited data transfer procedures

To permit the transfer of very urgent user data, transport protocols of classes 1, 2, 3 and 4 offer the option of an expedited data transfer service which routes information outside, and with higher priority than, the normal data flow. This service is reserved for sending very short messages which must not exceed 16 bytes of user data; it must not be confused with the priority mechanism for sending normal data which is optionally invoked at the time of establishment of the transport connection.

The expedited data is submitted to the transport entity as a parameter of a TEXreq primitive (Figure 3.56). The data is sent by an ED TPDU (*expedited data*) whose structure is the same as that of a DT TPDU (Figure 3.55). The message end bit is always set to 1, since the expedited data is considered to consist of isolated blocks which cannot be segmented. At the destination entity, receipt of the ED TPDU causes expedited data to be sent to the distant user as a parameter of a TEXind primitive.

In order to permit flow control and possibly recovery on error, the ED data units are numbered. The flow of expedited data units is clearly different from that of normal data units; this imposes a distinct sequence numbering on the flow of the latter. This leads to acknowledgement of ED TPDUs by special EA data units (*expedited data acknowledgement*). These EA TPDUs have the same structure as the AK TPDU (Figure 3.55) with the sole exception of the credit field CDT which is absent in this case.

The sending entity transmits the expedited data by placing the corresponding ED TPDU at the head of the TPDU transmit queue on the required network connection. The ED TPDU is then passed to the network by an NDTreq normal network data transfer primitive (Figure 3.57). It can be seen that the problem of

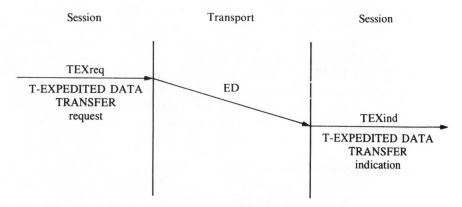

Figure 3.56 Expedited data transfer

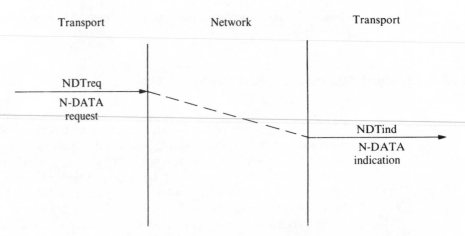

Figure 3.57 Normal network data transfer

priority routing of expedited data is taken care of by the transport entity independently from the network layer.

The expedited data transfer service guarantees that the expedited data is delivered to the destination before all normal data which has been sent after it. It can be seen immediately that the preceding mechanism allows this condition to be satisfied with classes 1, 2 and 3, since they assume that the network maintains the transferred TPDUs in sequence. This is no longer the case with class 4 which can use a nonsequencing network. To achieve priority of delivery in spite of the nonsequencing, it is necessary in class 4 to suspend transmission of data units after sending an ED TPDU until receipt of the corresponding EA TPDU. This technique thus leads to the adoption of a policy such that each ED TPDU is acknowledged individually and where only one TPDU can be in transit on one transport connection at a given time. It can be seen that the credit mechanism cannot be applied here; this explains the absence of the CDT field in the EA TPDU for acknowledging expedited data. As far as the receiver buffers are concerned, the only requirement is that the transport entity always saves an ED receive buffer for each open transport connection with the expedited option. This ensures that a transport entity is always capable of receiving an ED TPDU.

With the class 1 protocol, it is possible to use an alternative expedited data transport service which uses the network expedited data service if it exists. In this case, the ED and EA TPDUs are sent to the network as parameters of an N-EXPEDITED DATA request (NEXreq) primitive instead of an NDTreq normal data primitive (Figure 3.58). This means that the transport entity no longer has to manipulate the normal data queue to send the expedited data in priority, since this function is taken care of by the network.

3.3.8 Processing protocol errors

The procedure for processing protocol errors is operated in all classes following receipt of an invalid TPDU. Following this event, the transport entity may ignore

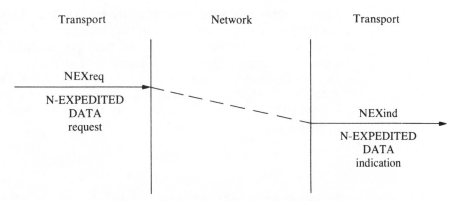

Figure 3.58 Expedited network data transfer

the TPDU, but more often it must undertake an action specific to the failure in order to preserve the connections not yet affected by the failure. These actions can, for example, consist of releasing the transport connection, closing the network connection or resetting the network connection. The transport entity can also send to its counterpart an error ER TPDU of which one field can contain the header of the invalid TPDU. The entity which sent the invalid TPDU can thus initiate the relevant recovery procedure.

3.3.9 Checksum

The checksum is a 16-bit error check word which can be inserted in the TPDU for error detection in class 4. This check word could be calculated from a polynomial code, but this choice is not justified in this case since the errors which affect the TPDU do not necessarily occur in bursts. Moreover, the polynomial code used in the TPDU could be redundant with the polynomial code which may be used in the data link protocol. Finally, the transport protocol is usually implemented by software and it is thus necessary that error detection should be performed efficiently with software; this is not the case with polynomial codes which are better suited to hardware implementation.

Error detection in the ISO transport layer is provided by inserting a *checksum* consisting of two consecutive bytes into the TPDU. These are computed in such a way as to make the modulo 255 sum and the weighted sum modulo 255 of all the bytes of the TPDU including the two check bytes equal to zero, hence

$$\sum_{i=1}^{L} a_i \equiv 0 \quad \text{modulo } 255 \tag{3.1}$$

$$\sum_{i=1}^{L} i\, a_i \equiv 0 \quad \text{modulo } 255 \tag{3.2}$$

where i is the index of the byte within the TPDU, a_i is the value of the byte in position i and L is the length of the TPDU expressed as a number of bytes.

This algorithm may seem difficult to implement since the weighted sum (3.2) depends on the indices n and $n + 1$ of the check bytes whose position varies with the type and format of the TPDU. In fact this does not matter since the standard suggests a simple algorithm for computing the checksum.

This approach, which is represented in the flowchart of Figure 3.59, consists of recursively computing two auxiliary variables $C0$ and $C1$. The two variables are initialized to zero and for each byte of the TPDU, the following operations are performed:

$$C0 := C0 + a_i \qquad (3.3)$$

$$C1 := C1 + C0 \qquad (3.4)$$

As the values of the two bytes a_n and a_{n+1} of the checksum are not yet evaluated, the corresponding operations (3.3) and (3.4) are bypassed. After computation of Equations (3.3) and (3.4) relating to the last byte of the TPDU, the two bytes of the checksum are determined from the respective equations:

$$a_n = -C1 + (L - n)\,C0 \qquad (3.5)$$

$$a_{n+1} = C1 - (L - n + 1)\,C0 \qquad (3.6)$$

To check that Equation (3.1) is indeed enforced, it can be noted that computation of $C0$ using (3.3) on the whole TPDU produces a value of $C0$ such that

$$\sum_{i=1}^{L} a_i \equiv C0 + a_n + a_{n+1} \qquad (3.7)$$

Using (3.5) and (3.6) it is clear that

$$a_n + a_{n+1} = -C0 \qquad (3.8)$$

which shows that Equation (3.1) is satisfied.

Now consider Equation (3.2). The algorithm computed by (3.4) on the whole TPDU gives a value of $C1$ such that

$$\sum_{i=1}^{L} (L - i + 1)a_i \equiv C1 + (L - n + 1)a_n + (L - n)a_{n+1} \qquad (3.9)$$

By substituting the values of a_n and a_{n+1} defined by (3.5) and (3.6) into (3.9), the following is obtained:

$$\sum_{i=1}^{L} (L - i + 1)a_i \equiv C1 + (L - n + 1)[-C1 + (L - n)C0]$$
$$+ (L - n)[C1 - (L - n + 1)C0] \qquad (3.10)$$

or

$$\sum_{i=1}^{L} (L - i + 1)a_i \equiv 0 \quad \text{modulo } 255 \qquad (3.11)$$

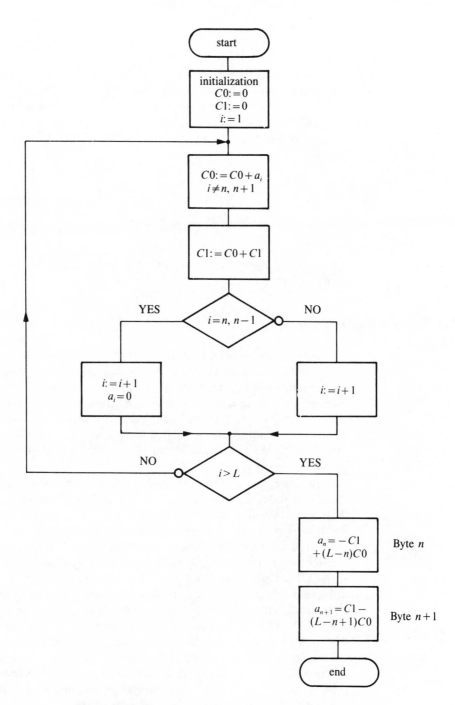

Figure 3.59 Flowchart of the checksum generation algorithm

Equation (3.11) is identical to

$$\sum_{i=1}^{L} (L - i + 1)a_i \equiv (L + 1) \sum_{i=1}^{L} a_i - \sum_{i=1}^{L} ia_i \equiv 0 \qquad (3.12)$$

It can be seen that Equation (3.11) implies that Equation (3.2) is valid if (3.1) is also valid, which is indeed the case.

At the receiver, the algorithm is organized in the same manner, with the sole difference that the checksum bytes are included in the computation of $C0$ and $C1$ (Figure 3.60). After the last byte of the TPDU, the values of $C0$ and $C1$ are thus given by

$$C0 \equiv \sum_{i=1}^{L} a_i \quad \text{modulo } 255 \qquad (3.13)$$

$$C1 \equiv \sum_{i=1}^{L} (L - i + 1)a_i \quad \text{modulo } 255 \qquad (3.14)$$

$C0$ and $C1$ must, therefore, be zero and a value different from zero detects transmission errors.

3.3.10 Class 0 ISO transport layer

Operation of the various ISO transport protocol classes is defined by state tables. In practice, it is often convenient to make use of state diagram representation, for example by using the symbols of the SDL language. In order to illustrate this approach, the state diagrams relating to class 0 are shown in Figures 3.61–3.65.

The transport connection can take five different states which are *open* (connection established), *closed* (connection released), WFNC (waiting for network connection), WFCC (waiting for CC TPDU) and WFTRESP (waiting for TCONrsp primitive).

In the case, for example, of connection establishment, the latter is initially in the *closed* state. Operations start with a TCONreq connection request primitive (Figure 3.61) and if this is acceptable and no network connection is available, the transport entity requests establishment of a network connection with an NCONreq primitive and then passes to the WFNC state.

Establishment of the network connection is signalled to the transport entity by an NCONcnf primitive; this permits a CR TPDU transport connection request to be sent (Figure 3.62). The connection then passes to the WFCC state.

At the distant transport entity, the connection is still in the *closed* state. Receipt of the CR TPDU causes a TCONind primitive to be sent to the user to indicate the connection request. The system then passes to the WFTRESP state while waiting for a response to the connection request (Figure 3.61).

When the distant transport entity receives the TCONrsp primitive from its user, it can switch to the *open* state after sending a CC TPDU to confirm the connection (Figure 3.65).

At the other end of the connection, receipt of the CC TPDU causes a TCONcnf connection confirmation primitive to be sent to the local user and a transition of the connection to the *open* state (Figure 3.63).

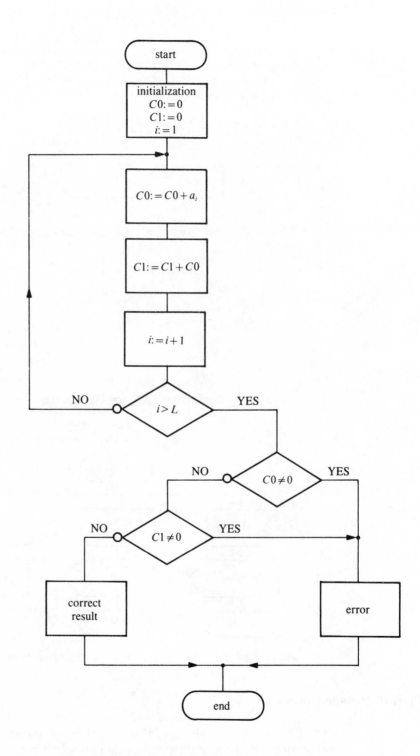

Figure 3.60 Flowchart of the checksum verification algorithm

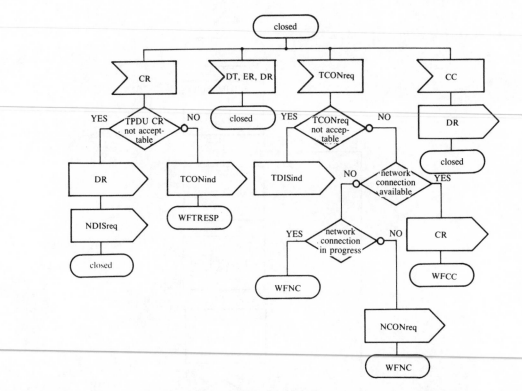

Figure 3.61 Transport protocol class 0. Operation from the *closed* state

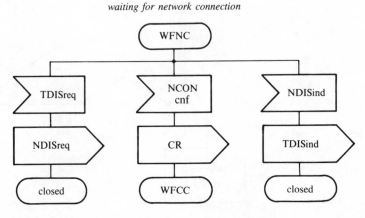

Figure 3.62 Transport protocol class 0. Operation from the WFNC state

3.3.11 NBS transport protocol

The NBS transport protocol [3.15] defines class 2 and class 4 protocols which are almost identical to the corresponding ISO protocol classes. The main differences in 1984 were as follows:

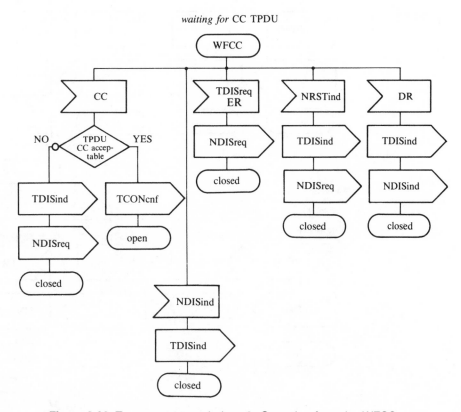

Figure 3.63 Transport protocol class 0. Operation from the WFCC state

(a) The NBS protocol provides an optional procedure for graceful close of the transport connection
(b) The use of extended sequence numbering is mandatory with the NBS protocol while it is optional with the ISO standard
(c) The NBS protocol provides a fast select message transfer procedure
(d) The use of explicit flow control is mandatory in class 2 with the NBS protocol

3.4 CONNECTIONLESS TRANSPORT PROTOCOLS

3.4.1 Overview of connectionless transport protocols

It has been seen that to ensure reliable end-to-end transport, it is necessary to operate relatively complex sequencing, flow control and error-detection functions. These functions are difficult to provide with a *connectionless transport service* so the latter is restricted to applications which require the shortest possible response time and which can accommodate occasional errors in messages.

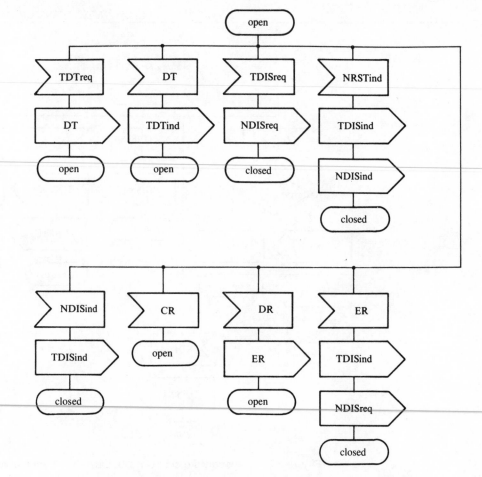

Figure 3.64 Transport protocol class 0. Operation from the *open* state

waiting for TCONrsp

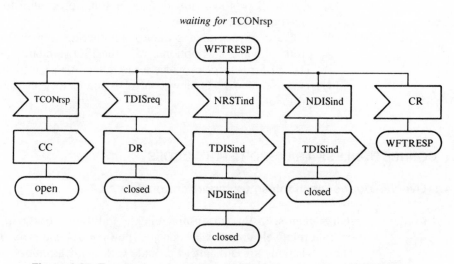

Figure 3.65 Transport protocol class 0. Operation from the WFTRESP state

Connectionless transport can be performed according to two different techniques which are the counterpart of the two solutions proposed for the connectionless mode in the network layer.

The first approach is to alter the connection oriented transport protocol to transfer a short message at the time of connection establishment. This message is delivered to the destination and the connection is immediately aborted in the same way as with the *fast select* option proposed for the X.25 network service. The NBS transport protocol offers a service of this type [3.15].

The ISO is carrying out standardization of a connectionless transport protocol [3.18] based on a different technique with a service similar to the *datagram* service. This technique has the advantage over the previous one of being more efficient for the transmission of short messages, but it requires a fundamentally different protocol from that used for the mode with connection. In contrast, the transfer of single messages by the fast select service is useful in the sense that it can be provided by a relatively minor modification of the basic connection mode protocol.

3.4.2 Transfer of single messages with the NBS transport protocol

The NBS protocol uses three additional primitives for the connectionless service; these are TUNIreq, TUNIind and TUNIcnf.

Transfer of an isolated message, whose size does exceed 32 bytes, is performed in this case without previously establishing a transport connection. To send an isolated message, the transport layer user sends a T-UNIT-DATA request primitive (TUNIreq) which contains the data to be transmitted as a parameter. Unlike the normal data transfer request TDTreq, the TUNIreq primitive also contains the same parameters as a connection request, particularly the calling and called addresses, together with an indication of the required quality of service.

The transport entity reads the TUNIreq primitive as a specific connection request which causes it to send a CR TPDU connection request to its counterpart; the primitive contains the user data with the special feature that a parameter of the data unit indicates that it consists of a single message (Figure 3.66). In class 2, and hence with a reliable network, receipt of the CR TPDU by the destination transport entity causes the message data to be sent to the local user by a T-UNIT-DATA indication primitive (TUNIind). The called transport entity then aborts the connection after sending a DR TPDU disconnect request to the sender. At the sender, receipt of the DR TPDU allows checking that the transfer has been performed, clearing of the connection and notification of the result of the operation to the user by a T-UNIT-DATA confirm primitive (TUNIcnf).

The same procedure can be applied in class 4, but as the network connection in this case has a non-negligible error rate, it is possible to use the transfer option indicated in Figure 3.67. In this case, the data is not passed immediately to the user at the time of receipt of the CR TPDU since its integrity must be verified. To do this, the remote entity can return the data with the disconnect data unit DR. The sending entity can then check that the data which it has transmitted is the same as that which has been returned to it; this permits it to acknowledge

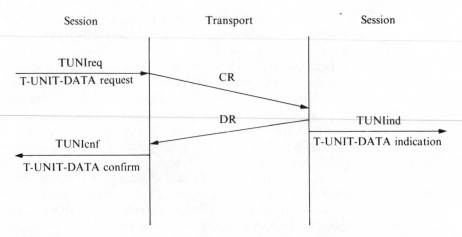

Figure 3.66 Transfer of a single short message on an aborted connected request

the transfer by sending a DC TPDU. After receipt of the DC TPDU, the remote transport entity passes the data to the local user with a TUNIind primitive.

When the isolated message to be transferred exceeds 32 bytes, the previous procedures are no longer applicable. The message must then be sent by a DT TPDU, or possibly several if segmentation is necessary (Figure 3.68). The disconnect DR TPDU can no longer contain a replica of the transmitted data and it can carry only one error checking indication which conditions the sending of the disconnect confirm TPDU.

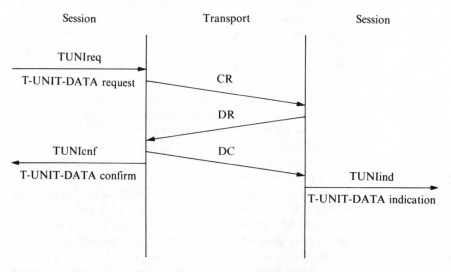

Figure 3.67 Transfer of a short single message on an aborted connection request and confirmation of disconnection

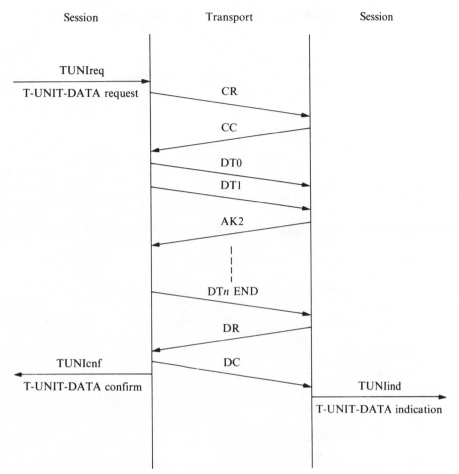

Figure 3.68 Transfer of a single long message on an aborted connection request and confirmation of disconnection

3.4.3 ISO DIS 8602 connectionless transport protocol

The connectionless ISO transport protocol provides services in the session layer by means of only two primitives TUNIreq and TUNIind which have practically the same parameters as the corresponding primitives of the NBS protocol (Table 3.8).

The transport layer uses the services of the network layer which in this case can be with or without connection. In the first case, only the NCONreq, NCONind, NDTreq, NDTind, NRSTreq, NRSTind, NRSTrsp, NRSTcnf, NDISreq and NDISind primitives are used since the connect confirm, acknowledgement and expedited data services are not of use. In the case of a connectionless network layer, the only network service primitives are NUNIreq and NUNIind which are indicated in Table 3.9.

Table 3.8 Connectionless ISO transport service primitives.

Symbol	Primitive	Parameters
TUNIreq	T-UNITDATA request	Source address Destination address Quality of service User data (message)
TUNIind	T-UNITDATA indication	Source address Destination address Quality of service User data (message)

Table 3.9 Connectionless ISO network service primitives.

Symbol	Primitive	Parameters
NUNIreq	N-UNITDATA request	Destination address Source address Quality of service User data
NUNIind	N-UNITDATA indication	Destination address Source address User data

The functions provided by the connectionless transport layer are as follows:

(a) Selection of the network service best suited to the required quality of service
(b) Mapping of the network and transport addresses
(c) Delimiting of the start and end of a TSDU
(d) Error detection

When the transport layer uses a connectionless network, the transport protocol provides only a single type of data unit, the UD TPDU for isolated data transfer. In its fixed part this data unit contains only the length byte LI and the TPDU type identification byte UD. The variable part contains the source and destination addresses of the transport service access points TSAP, together with the checksum for error detection (Figure 3.69). Clearly there is no *reference* since the transfers are performed without establishment of a connection.

Transfer of an isolated message starts on the initiative of one of the users which passes the data by means of a TUNIreq primitive to the corresponding transport entity (Figure 3.70). The latter checks whether the length of the data block

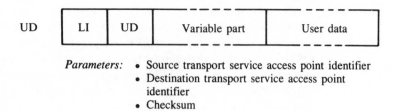

| UD | LI | UD | Variable part | User data |

Parameters: • Source transport service access point identifier
• Destination transport service access point identifier
• Checksum

Figure 3.69 Structure of the ISO connectionless UD TPDU transport protocol

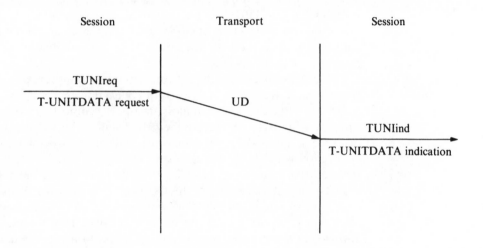

Figure 3.70 Transfer of a single message with the connectionless ISO transport protocol

submitted to it is compatible with the maximum packet size on the network. If this condition is not satisfied, the transfer request is rejected and the abort is signalled locally to the user. In the other case, the transport entity decides from the required quality of service whether or not it is necessary to insert a checksum in the UD TPDU. Also, it determines the identification of the source and destination network service access points from the transport address parameters of the TUNIreq primitive. After these operations, the transport entity sends the message by putting it in the user data field of a UD TPDU which it sends to the network layer by an NUNIreq primitive. At the remote transport entity, the integrity of the TPDU is checked by means of the checksum, if it exists, and the UD TPDU is discarded if it contains errors. Otherwise, the data is passed to the local user by a TUNIind primitive.

When the transport layer uses a network with connection, the operations start, as previously after a TUNIreq primitive, with mapping of the transport and network addresses and a decision concerning the use of a checksum in the UD TPDU. If a network connection is already established, the transport entity immediately sends a UD TPDU which contains the user data. When this condition is not satisfied, the transport entity must request establishment of a network connection with an NCONreq primitive and send the UD TPDU only after receiving confirmation of the connection by an NCONcnf primitive. The network connection establishment request must contain a special data unit as a user data parameter; this special unit is called a UN TPDU and is sent to the distant transport entity to indicate to it that this network connection is reserved for a connectionless transport service.

At the destination transport entity, establishment of a network connection intended to service isolated messages is indicated by an NCONind primitive which contains the UN TPDU. Once this connection is established, UD TPDUs which arrive on this connection are treated as for the connectionless network service with data sent to the local user by a TUNIind primitive after possible checking by means of a checksum.

The checksum used with the connectionless transport service is the same as that which is used in the service with connection (§ 3.3.9).

Operation of the connectionless transport protocol can be defined from only three possible states; these are the *closed* state, without network connection, the *wait* for a network connection state and the *active* state where a network connection is established (Figures 3.71–3.73).

When the network operates in connectionless mode, the protocol can clearly only be in the *closed* state. In the case of a network with connection, it is necessary to release the network connection if it remains unused for too long, since costly resources must not be kept inactive. This is achieved by means of a **transacting timer** whose guard time TT is set to a value equal to the maximum delay in sending an isolated message. This timer is set or reset each time a new UD TPDU is sent and the network connection is released on expiry of the guard time.

3.5 IMPLEMENTATION OF A TRANSPORT LAYER

3.5.1 Problems in implementing a transport protocol

The standards which define communication protocols specify the mechanisms of the protocol, the formats and the interfaces in a conceptual manner but leave the choice of a particular implementation to the designer. The ISO and NBS standards follow this rule and leave considerable latitude of choice which opens the door to appreciable performance differences or even incompatibilities between two different implementations of the same protocol [3.15]. The main problems which arise in implementing a transport protocol in connection mode will be reviewed briefly.

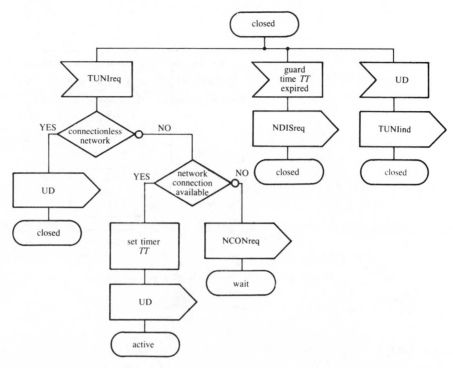

Figure 3.71 Connectionless ISO transport protocol. Operation from the *closed* state

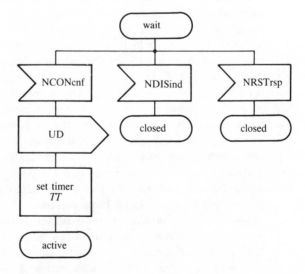

Figure 3.72 Connectionless ISO transport protocol. Operation from the *waiting* state of a state *wait for a network connection*

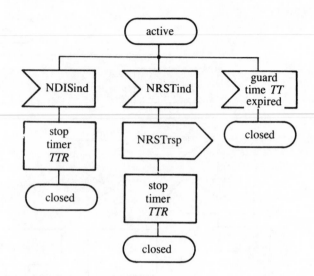

Figure 3.73 Connectionless ISO transport protocol. Operation from the *active* state

3.5.2 Timers

A complete class 4 transport layer can contain up to a dozen different types of timer. Setting the guard times for each type of timer is a difficult problem since the choice which must be made depends on the characteristics of the network and can have a direct impact on the system performance.

In the case of the *retransmission timer*, the guard time $T1$ defines the time which elapses between sending a DT TPDU and its replica when an acknowledgement has not been received in the meantime. In order to avoid the waste of resources which would result from needless retransmission, the guard time $T1$ must be set to a value which exceeds the sum of the outward transfer time of the DT TPDU, the processing time at the remote entity and the inward transfer time of the AK TPDU. The processing time can vary greatly as a function of the throughput on the other transport connections and the load corresponding to other functions provided by the equipment which executes the transport protocol. Moreover, the DT TPDU transfer time increases with TPDU size and varies considerably as a function of the traffic on the network.

It can be seen that the delay in acknowledging a DT TPDU has a mean value which depends greatly on the network (local, long haul, satellite links) but is generally long and has a large variance. In a long-haul network, mean transfer times of 100 to 200 ms can be expected with a processing time of the order of 100 ms; this leads to a mean acknowledgement delay of the order of 300–500 ms. Under these conditions, it is not possible to set the value of the guard time $T1$ once and for all since this would lead to a choice of value which would be too high under most circumstances with the consequence of an excessive retransmission delay and a corresponding performance degradation.

The commonest solution to the problem of the retransmission timer is to define the guard time $T1$ as a parameter which is set at installation and later adjusted by the network management service. A good compromise between speed of retransmission and economy of resources is to take a value for $T1$ which is of the order of double the mean acknowledgement delay while accepting a certain unnecessary retransmission rate. Another possible approach is to adjust the guard time $T1$ at the time of establishing the connection, or even to adopt a completely adaptive method, where the guard time is adjusted continuously as a function of the network transfer delay. The protocol does not provide dedicated mechanisms for this type of operation, but it is possible, for example, to measure the acknowledgement delay and compute a $T1$ value from it. This approach is attractive but it raises numerous problems which often make it impracticable. If some TPDUs may be duplicated in a network without sequencing, it is not always possible to determine the acknowledgement corresponding to a given TPDU. Also, the automatic retransmission delay control system must operate on mean values to avoid instability. As the transfer time can vary abruptly, for example following sudden congestion, the system could adjust itself on the basis of outdated information. In most cases this leads to selecting a fixed value for $T1$ which is determined at the time of installation or connection establishment.

When the acknowledgement of a TPDU is not received before expiry of the guard time $T1$, the transport entity must retransmit and the process can be repeated up to a maximum of N retransmissions. During the last retransmission, the transport entity must initiate a **give-up timer** and clear the connection if no acknowledgement is received before expiry of the corresponding guard time TGT. In practice, the N transmissions of a TPDU do not represent N independent attempts if they have all taken place during temporary congestion of the network. It is therefore necessary for the guard time TGT to open a time window whose duration corresponds to the *maximum acknowledgement delay* (and no longer the mean) relating to the last retransmitted TPDU. This generally leads to the choice of a very high value for TGT.

As an example, Table 3.10 shows typical values of guard times for a class 4 transport protocol using an X.25 network with terrestrial links.

3.5.3 Acknowledgement control policy

The acknowledgement data units AK serve the purpose of sequencing, error correction and flow control according to the particular modes in each protocol class. The standard leaves a number of choices open which can have a significant effect on the efficiency of resource utilization, the throughput and the response time. Transmission of AK TPDUs consumes resources and it is clearly desirable to avoid sending acknowledgements at a higher rate than is strictly necessary. Also, the choice of the source of the acknowledgement and the credit allocation policy can have a large impact on the system performance.

In connection with the rate of sending AK TPDUs, the simplest solution is to return an acknowledgement for each DT data unit received. This technique is clearly robust, but it wastes resources since it leads to transmission of more

Table 3.10 Typical guard times of some class 4 transport protocol timers using an X.25 network with terrestrial links.

Timer	Function	Guard time (seconds)
Retransmission ($T1$)	Acknowledgement delay	1
Inactivity	Time between receipt of two consecutive TPDUs	6
Window	Maximum delay between two window updates	2
Give-up timer (TGT)	Connection abort delay after N unsuccessful retransmissions	75
Network connection activity monitoring	Maximum delay for retaining an unused network connection	50
Frozen reference	Duration of reference freezing	90

acknowledgements than are strictly necessary. The efficiency of the procedure can be improved by concatenating the AK TPDU with another TPDU if one exists on the same network connection at the time of transmission.

A more efficient, but more difficult to implement, technique is to acknowledge several DT TPDUs with the same AK TPDU. The problem with this is to determine the time during which sending of an acknowledgement is suspended while waiting for the arrival of other DT TPDUs. Excessive retention of the acknowledgement by the destination could lead to expiry of transmission credit at the sender of the DT TPDUs and hence a throughput reduction on the transport connection. A good policy is to consider that a DT TPDU must normally be followed after a brief deadtime by other DT TPDUs if it is not the last of a message. This leads to suspending DT TPDU acknowledgements which do not carry the END of message mark in order to send a single AK TPDU for several DT TPDUs. It is then necessary to monitor the retention time of the acknowledgement with a timer which is started during receipt of a TPDU without an END mark and to send an AK TPDU as soon as the guard time runs out or a TPDU carrying the END of message mark arrives.

Another important decision left to the designer is that of the source of acknowledgements. One possible approach is for the transport layer to ensure that the transport user has correctly received the data before sending the corresponding AK TPDU. This technique is very reliable, since it offers the user the facility of discarding a DT TPDU if it cannot accept it, with the guarantee that this TPDU will be retransmitted if the protocol contains a retransmission mechanism, that is if it is of class 4. In contrast to this advantage, the time to produce the acknowledgement is in this case longer than for the case when sending the AK TPDU is done directly by the transport entity; this can reduce the maximum throughput on the transport connection.

In most cases, the transport entity itself assumes responsibility for sending the acknowledgements. This solution is more logical, since the protocol layers are supposed to be independent. The transport entity can also time the acknowledgements as a function of the overall state of the transport layer; this enables it, for example, to favour urgent traffic by reducing the rate on lower priority connections. The transport entity must clearly save all the acknowledged TPDUs until they are accepted by the session entity; this can pose problems if the latter is in a state of temporary congestion.

3.5.4 Flow control

Flow control is provided in the ISO and NBS protocols by a credit mechanism where the destination provides the sender of the DT TPDUs with an allocation to transmit a number of DT data units defined by the CDT field of an AK TPDU. The credit corresponds exactly to the number of DT TPDUs which the destination is ready to receive, starting from the last DT TPDU which it acknowledged with the AK TPDU.

The ISO and NBS standards leave the choice of credit allocation policy to the designer. The simplest control technique is to reserve a number of receive buffers equal to the agreed credit; this guarantees that the received DT TPDUs can always be processed. This method is thus very robust but it has the disadvantage of poor use of the buffers since these are not dynamically shared. Also, it tends to limit the maximum throughput on the transport connections since the destination can only grant credits in accordance with the exact number of free buffers at the time of allocation.

The policy of pre-allocating buffers as a function of the granted credits is very pessimistic in most cases. The transfer times of the TPDUs are generally long, except in the case of local area networks, so that some of the reserved buffers have every chance of being free before expiry of the credit. Under these conditions, it is logical to adopt an optimistic policy for long-haul networks where the destination grants more credits than the number of buffers available at a given time and allocates a buffer to a connection only at the time of arrival of a DT TPDU. With this approach, some DT data units can be lost if there is a lack of free buffers when they arrive at the destination. This method is thus applicable only if there is a retransmission mechanism for lost TPDUs; this corresponds to a class 4 protocol. Also, it is clearly necessary to prevent the loss of too many DT TPDUs due to lack of free buffers. One possible solution is to provide an adaptive mechanism such that the destination assigns more and more credits until it has to reject DT TPDUs following a lack of free buffers. This defines a limit which must not be exceeded and enables the credit allocation to be adjusted to a reasonable value. The procedure can be improved by making use of the option to reduce the upper window edge which is available in class 4. It is thus possible to slow down or block the sending of DT TPDUs for which a transmission credit has already been granted in order to cure congestion at the destination more rapidly. This operation clearly leads to the loss of DT TPDUs in transit; this is unimportant since they will be retransmitted and they could not in any case be received, since there are no free buffers. Operation of this procedure implies the use of flow control confirmation and sub-sequence numbering.

3.5.5 Local flow control

The transport protocol does not specify flow control at the interface between the transport layer and the session layer since this has a purely local significance and is therefore outside the OSI model. In practice, the designer must provide flow control at this interface in order to limit the flow of data submitted to the transport layer or sent to the session layer. This can be realized, for example, by a mailbox mechanism.

3.5.6 Concatenation

Concatenation permits the network to be used more efficiently by grouping several TPDUs into the same packet. The grouping rule states that the packet must contain only a single TPDU which contains user data but does not fix any other limit on the number of TPDUs which can be concatenated. If several transport connections are multiplexed onto the same network connection, the packet can concatenate the TPDUs from different connections. At the destination, it is convenient to process the NSDU carried by the packet in such a way as to simulate an NDTind primitive for each of the TPDUs which are concatenated.

3.5.7 Addressing

The transport addresses are those of the transport service user. They generally consist of the corresponding network address followed by a suffix specific to the transport service user. The network address serves to localize the transport entity.

Network addresses are generally global and must therefore be controlled by the network administrator. The transport suffix can have a purely local significance which permits it to be controlled at the user service level. To distinguish the various transport connections which can be used by the same user, transport entities assign a unique *reference*, which has a purely local significance, to each connection.

The connection request data unit CR contains the *source reference* together with the source and destination transport addresses if these are not identical to the corresponding network addresses, that is if there are several users of the network connection service. These transport addresses contained in the CR TPDU do not need to be complete and it is sufficient if they contain only the transport suffix. The destination transport entity knows the network source and destination addresses of the connection on which it receives the CR TPDU.

Even when the transport suffixes are controlled locally, it is useful if some of them have a global significance in order to permit access to well-known generic services without requiring complex searches in a directory. This corresponds to the 'well-known address' concept for public utility services.

The *references* of a transport connection are allocated by the transport entities and are initially unknown to the two users of the connection. It is therefore necessary to provide a mechanism which permits each of the two users of the connection to identify the references when required for data transfer. Here again, this is a local procedure which is not specified by the standard. Exchanges can, for example, be organized according to a method proposed by INTEL for its INA 960 transport software which is briefly described here [3.19].

With the INA 960 software, each of the two users of the transport layer must first signal its request to open a new virtual connection by sending an OPEN primitive. The transport software in return provides each user with the identifier of a transport connection descriptor which will then serve to identify the connection.

In the second stage, one of the users must signal with an AWAIT$CONNEC-TION$REQUEST primitive that it is ready to accept a connection corresponding to the identifier with which it has been provided. The user can specify with this primitive the type of connection which it is ready to accept with this identifier, for example a connection from any user or, in contrast, a connection with a particular user. The establishment of the connection itself is performed on the initiative of one of the two users, by a SEND$CONNECTION$REQUEST primitive. This primitive contains, as a parameter, the connection descriptor identifier, which has been provided in response to the OPEN primitive, together with the transport address of the distant user. When the latter has previously sent an AWAIT$CONNECTION$REQUEST primitive, the connection can be established and it is definitely represented by the two corresponding identifiers.

Data transfers can then take place, with transmission of data by a SEND$DATA primitive which points to the transmit data buffer and contains the connection identifier as a parameter.

At the other end, the user process must take the initiative of signalling to the transport software that it is ready to receive the data by allocating a number of receive buffers to it with a RECEIVE$DATA primitive which contains the connection identifier as a parameter. It can be seen that local flow control in this case is controlled by the user on the receiving side as well as the transmission side.

Closing of the connection is performed on the initiative of one of the users who sends a CLOSE primitive which has the effect of freeing the corresponding descriptors.

3.5.8 Examples of transport software

The INTEL INA 960 transport software implements the complete ISO 8073 transport protocol, including class 4 [3.19]. The memory size required for the basic service is of the order of 32 000 bytes and in practice it reaches around 64 000 bytes; this takes account of the auxiliary services and the space reserved for buffers and transport connection descriptors. With a dedicated INTEL 80186 microprocessor, the maximum transport layer throughput is of the order of 200 kbyte/s.

With another example of class 4 transport software realization developed at the *National Bureau of Standards*, the size of the program is of the order of 10 000 lines of code in the C language; this gives a result comparable to that of the previous example [3.20].

REFERENCES

CHAPTER 1

[1.1] M. Schwartz, *Computer Communication Network Design and Analysis*, Prentice-Hall, Englewood Cliffs, 1977.

[1.2] L. Kleinrock, *Queuing Systems. Vol. 1: Theory, Vol. 2: Computer Applications*, Wiley-Interscience, New York 1975, 1976.

[1.3] R. Eichaker and G. Barruel, *Temps de réponse des réseaux informatiques*, Masson, Paris, 1984.

[1.4] T. Saaty, *Elements of Queuing Theory*, McGraw-Hill, New York, 1961.

[1.5] E. Gelenbe and I. Mitrani, *Analysis and Synthesis of Computer Systems*, Academic Press, London, 1980.

[1.6] C. Macchi and J.F. Guilbert, *Téléinformatique*, Dunod, Paris, 1979.

[1.7] G. Pujolle, D. Seret, D. Dromard and E. Horlait, *Réseaux et télématique*, Eyrolles, Paris, 1985.

[1.8] V. Ahuja, *Design and Analysis of Computer Communication Networks*, McGraw-Hill, New York, 1982.

[1.9] M. Rudnianski, *Architecture de réseaux: le modèle ISO*, Editests, Paris, 1986.

[1.10] H. M. Wagner, *Principles of Operations Research*, Prentice-Hall, Englewood Cliffs, 1969.

[1.11] M. Schwartz, *Telecommunication Networks, Protocols, Modeling and Analysis*, Addison Wesley, Reading, 1987.

[1.12] M. Reiser, Performance evaluation of data communication systems, *Proceedings of the IEEE*, Vol. 70, No. 2, pp. 171–196, February 1982.

[1.13] H. Inose and T. Saito, Theoretical aspects in the analysis and synthesis of packet communication networks, *Proceedings of the IEEE*, Vol. 66, No. 11, pp. 1409–1422, November 1978.

[1.14] F. Tobagi, M. Gerla, R. Peebles and E. Manning, Modeling and measurement techniques in packet communication networks, *Proceedings of the IEEE*, Vol. 66, No. 11, pp. 1423–1447, November 1978.

[1.15] J.R. Jackson, Two shop-like queuing systems, *Management Science*, Vol. 10, pp. 131–142, 1963.

[1.16] D. Towsley and J. Wolf, On the statistical analysis of queue lengths and waiting times for statistical multiplexers with ARQ retransmission schemes. *IEEE Transactions on Communications*, Vol. COM-27, No. 4, pp. 693–702, April 1979.

[1.17] M. Easton, Batch throughput efficiency of DCCP/HDLC/SDLC selective reject protocols, *IEEE Transactions on Communications*, Vol. COM-28, No. 2, pp. 187–195, February 1980.

[1.18] J. Labetoulle and G. Pujolle, HDLC throughput and response time for bidirectional data flow with nonuniform frame sizes, *IEEE Transactions on Computers*, Vol. C-30, No 6, pp. 405–413, June 1981.

[1.19] M. Reed, and T.D. Smetanka, Implications of a selective acknowledgment scheme on satellite performance, *IBM Journal of Research and Development*, Vol. 23, No 2, pp. 189–196, March 1979.

[1.20] J. Wang, Delay and throughput analysis for computer communications with balanced HDLC procedures, *IEEE Transactions on Computers*, Vol. C-31, No 8, pp. 739–746, August 1982.

[1.21] C. Fujiwara, M. Kasahara, K. Yamashita and T. Namekawa, Evaluations of error control techniques in both independent-error and dependent-error channels, *IEEE Transactions on Communications*, Vol. COM-26, No 6, June 1978.

[1.22] L.W. Yu and J.C. Majithia, An analysis of one direction of window mechanism, *IEEE Transactions on Communications*, Vol. COM-27, No 5, pp. 778–788, May 1979.

[1.23] W.W. Chu and A.G. Konheim, On the analysis of a class of computer communication systems, *IEEE Transactions on Communications*, Vol. COM-20, No 3, pp. 645–660, June 1972.

[1.24] W.W. Chu, Buffer-behavior for batch Poisson arrivals and single constant output, *IEEE Transactions on Communication Technology*, Vol. COM-18, No 5, pp. 613–618, October 1970.

[1.25] H. Rudin, Performance of simple multiplexer concentrators for data communications, *IEEE Transactions on Communication Technology*, Vol. COM-19, No 2, pp 178–187, April 1971.

[1.26] B. Meister, H.R. Muller and H. Rudin, New optimization criteria for message switching networks, *IEEE Transactions on Communication Technology*, Vol. COM-19, No 3, pp. 256–260, June 1971.

[1.27] L. Kleinrock, Analytic and simulation methods in computer network design, *Proceedings AFIPS 1970 Spring Joint Computer Conference*, pp. 569–579, 1970.

[1.28] B. Meister, H.R. Muller and H. Rudin, On the optimization of message switching networks, *IEEE Transactions on Communications*, Vol. COM-20, No 1, pp. 8–14. February 1972.

[1.29] K. Maruyama and D.T. Tang, Discrete link capacity and priority assignments in communication networks, *IBM Journal of Research and Development*, pp. 254–263, May 1977.

[1.30] K. Maruyama, L. Fratta and D.T. Tang, Heuristic design algorithm for computer communication networks with different classes of packets, *IBM Journal of Research and Development*, pp. 360–369, July 1977.

[1.31] H. Frank and H.W. Chou, Topological optimization of computer networks, *Proceedings of the IEEE*, Vol. 60, No 11, pp. 1385–1397, November 1972.

[1.32] M. Gerla, H. Frank, W. Chou and J. Eckl, A cut saturation algorithm for topological of packet switched communication networks, *National Telecommunications Conference Record*, pp. 1074–1085, December 2–4, 1974.

[1.33] K.M. Chandy and R.A. Russel, The design of multipoint linkages in a teleprocessing tree network, *IEEE Transactions on Computers*, Vol. C-21, No 10, pp. 1062–1066, October 1972.

[1.34] R.C. Prim, Shortest connection networks and some generalizations, *The Bell System Technical Journal*, Vol. 36, pp. 1389–1401, November 1957.

[1.35] L.R. Esau and K.C. Williams, On teleprocessing system design, Part II, *IBM Systems Journal*, Vol. 5, No. 3, pp. 142–147, 1986.

CHAPTER 2

[2.1] M. Schwartz, *Computer Communication Network Design and Analysis*, Prentice-Hall, Englewood Cliffs, 1977.

[2.2] M. Schwartz and T.S. Stern, Routing techniques used in computer communication networks, *IEEE Transactions on Communications*, Vol. COM-28, No 4, pp. 539–552, April 1980.

[2.3] R.T. Prosser, Routing procedures in communication networks. Part I: random procedures. Part II: directory procedures, *IRE Transactions on Communications Systems*, Vol. CS-10, No 4, pp. 322–335, December 1962.

[2.4] G. Ludwig and R. Roy, Saturation routing network limits, *Proceedings of the IEEE*, Vol. 65, No 9, pp. 1353–1361, September 1977.

[2.5] C. Petitpierre, Meshed local computer networks, *IEEE Communication Magazine*, Vol. 22, No 8, pp. 36–40, August 1984.

[2.6] D.G. Cantor and M. Gerla, Optimal routing in a packet-switched computer network, *IEEE Transactions on Computers*, Vol. C-23, No 10, pp. 1062–1069, October 1974.

[2.7] L. Fratta, M. Gerla and L. Kleinrock, The flow deviation method: an approach to store and forward communication network design, *Networks*, John Wiley, New York, pp. 97–133, 1973.

[2.8] D.J. Silk, Routing doctrines and their implementation in message switching networks, *Proceedings of the IEEE*, Vol. 116, No 10, pp. 1631–1638, October 1969.

[2.9] H. M. Wagner, *Principles of Operations Research*, Prentice-Hall, Englewood Cliffs, 1969.

[2.10] G.L. Fultz and L. Kleinrock, Adaptive routing techniques for store-and forward-computer communication networks, *Proceedings of the International Conference on Communications*, pp. 39.1–7, June 14–16, 1971.

[2.11] P Baran, On distributed communication networks, *IEEE Transactions on Communication Systems*, Vol. CS-12, pp. 1–9, March 1964.

[2.12] B.W. Boehm and R.L. Mobley, Adaptive routing techniques for distributed communication systems, *IEEE Transactions on Communication Technology*, Vol. COM-17, No 3, pp. 340–349, June 1969.

[2.13] R.G. Gallager, A minimum delay routing algorithm using distributed computation. *IEEE Transactions on Communications*, Vol. COM-25, No 1, pp. 73–85, January 1977.

[2.14] T. Cegrell, A routing procedure for the TIDAS message-switching network, *IEEE Transactions on Communications*, Vol. COM-26, No 6, pp. 575–585, June 1975.

[2.15] P.M. Merlin and A. Segall, A failsafe distributed routing protocol, *IEEE Transactions on Communications*, Vol. COM-27, No 9, pp. 1280–1287, September 1979.

[2.16] L. Kleinrock and F. Kamoun, Hierarchical routing for large networks, *Computer Networks*, Vol. 1, pp. 155–174, January 1977.

[2.17] H. Rudin et al., Special issue on congestion control in computer networks, *IEEE Transactions on Communications*, Vol. COM-29, No 4, pp. 373–535, April 1981.

[2.18] D.W. Davies, The control of congestion in packet-switching networks, *IEEE Transactions on Communications*, Vol. COM-20, No 3, pp. 546–550, June 1972.

[2.19] M.C. Pennotti and M. Schwartz, Congestion control in store and forward tandem links, *IEEE Transactions on Communications*, Vol. COM-23, No 12, pp. 1434–1443, December 1975.

[2.20] S.S. Lam and Y.C. Luke Lien, Congestion control of packet communication networks by input buffer limits. A simulation study, *IEEE Transactions on Computers*, Vol. C-30, No 10, pp. 733–742, October 1981.

[2.21] L.E.N. Delbrouk, A unified approximate evaluation of congestion functions for smooth and peaky traffics, *IEEE Transactions on Communications*, Vol. COM-29, No 2, pp. 85–91, February 1981.

[2.22] D.W. Davies, The control of congestion in packet-switching networks, *IEEE Transactions on Communications*, Vol. COM-20, pp. 546–550, June 1972.

[2.23] W.L. Price, Simulation studies of an isarithmically controlled store and forward data communication network, *Proceedings of the IFIP Congress 74*, North-Holland: Amsterdam, pp. 151–154, 1974.

[2.24] J.C. Majithia, M. Irland, J.L. Grangé, N. Cohen and C. O'Donnell, Experiments in congestion control techniques, *Flow Control In Computer Networks*, J.L. Grangé and M. Gien, North-Holland, Amsterdam, pp. 211–234, 1979.

[2.25] M.I. Irland, Buffer management in a packet switch, *IEEE Transactions on Communications*, Vol. COM-26, No 3, pp. 328–337, March 1978.

[2.26] K.D. Gunther, Prevention of deadlocks in packet-switched data transport systems, *IEEE Transactions on Communications*, Vol. COM-29, No 4, pp. 512–524, April 1981.

[2.27] P.M. Merlin and P.J. Schweitzer, Deadlock avoidance in store-and-forward networks. I. Store and forward deadlock. II. Other deadlock types, *IEEE Transactions on Communications*, Vol. COM-28, No 3, pp. 345–360, March 1980.

[2.28] CCITT, *Data Communication Networks Services and Facilities, Terminal Equipments and Interfaces. Recommendations* X.1–X.29, CCITT Yellow Book, Vol. VIII, Fascicle VIII.2, Geneva, 1981.

[2.29] CCITT, *Data Communication Networks, Interfaces, Recommendations* X.20–X.32, CCITT Red Book, Vol. VIII, Fascicle VIII.3, Geneva, 1985.

[2.30] A. Rybczynski, X.25 interface and end-to-end virtual circuit service characteristics, *IEEE Transactions on Communications*, Vol. COM-28, No 4, pp. 500–509, April 1980.

[2.31] H.C. Folts, X.25 transaction-oriented features. Datagram and fast select, *IEEE Transactions on Communications*, Vol. COM-28, No 4, pp. 496–500, April 1980.

[2.32] C.R. Dhas and V.K. Konangi, X.25: an interface to public packet networks, *IEEE Communications Magazine*, Vol. 24, No 9, pp. 18–25, September 1986.

[2.33] K.L. Cohen and R.P. Levy, X.25 implementation. The untold story, *Proceedings SIGCOM 83 Symposium*, pp., 60–64, March 8–9, 1983.

[2.34] PROTOCOM, Sync equipment gets X.25 link, *Electronics*, p. 236, November 3, 1983.

[2.35] F.M. Burg and C.T. Chen, Of local networks, protocols and the OSI reference model, *Data Communications*, pp. 129–150, November 1984.

[2.36] PTT, *TELEPAC.X.28 Specification. Release 3*, Direction générale des PTT, Berne, 1985.

[2.37] M.K. Molloy, Character delays in simple X.3 PAD devices. *Proceedings SIGCOM 83*, pp. 275–279, March 8–9, 1983.

[2.38] CCITT, *Data Communication Networks. Open Systems Interconnection* (OSI). *System Description Techniques. Recommendations X.200–X.250*, Red Book, Vol. VIII, Fascicle VIII.5, CCITT, Geneva, 1985.

[2.39] ISO, *Information Processing Systems. Data Communications. Use of X.25 to Provide the OSI Connection-Mode Network Service*, Draft International Standard ISO/DIS 8878, ISO, Geneva, 1986.

[2.40] ISO, *Information Processing Systems. Data Communications. Internal Organization of the Network Layer*, Draft International Standard ISO/DIS 8648, ISO, Geneva, 1986.

[2.41] ISO, *Information Processing Systems. Data Communications. Protocol for Providing the Connectionless-Mode Network Service*, Draft International Standard ISO/DIS 8473, ISO, Geneva, 1985.

[2.42] ISO, *Information Processing Systems. Data Communications. Network Service Definition*, Draft International Standard ISO/DIS 8348, ISO, Geneva, 1985.

[2.43] ISO, *Information Processing Systems. Data Communications. Protocol for Providing the Connectionless-Mode Network Service. Addendum 1: Provision of the Underlying Service assumed by ISO 8473*, Draft International Standard ISO/DIS 8473/DAD 1, ISO, Geneva, 1986.

[2.44] MAP TASK FORCE, *Manufacturing Automation Protocol Specification*, General Motors, Warren, September 1984.

CHAPTER 3

[3.1] M. Rudnianski, *Architecture de réseaux: le modèle ISO*, Editests, Paris, 1986.

[3.2] A.S. Tanenbaum, *Computer Networks*, Prentice-Hall, Englewood Cliffs, 1981.

[3.3] W. Stallings, *Data and Computer Communications*, Macmillan, New York, 1985.

[3.4] W. Stallings, A primer: understanding transport protocols, *Data Communications*, pp. 389–397, November 1984.

[3.5] C.A. Sunshine, *Computer Communications: Architectures, Protocols and Standards*, IEEE, Piscataway, pp. 345–387,1985.

[3.6] K.G. Knightson, The transport layer standardization, *Proceedings of the IEEE*, Vol. 71, No 12, pp. 1394–1396, December 1983.

[3.7] P.F. Linington, The transport service proposed by PSS study group 3, *Online*, Northwood Hills, pp. 621–634, 1980.

[3.8] P. von Studnitz, Transport protocols; their performance and status in international standardization. *Computer Networks*, pp. 27–35, 1983.

[3.9] K.G. Knightson, The transport layer, *Proceedings of the Sixth International Conference on Computer Communication*, London, pp. 787–791, 7–10 September 1982.

[3.10] V.L. Voydock, Security mechanisms in a transport layer protocol, *Computer Networks*, pp. 433–449, 1984.

[3.11] C.A. Sunshine and Y.K. Dalal, Connection management in transport protocols, *Computer Networks*, pp. 454–473, 1978.

[3.12] D.P. Sidhu and T.P. Blumer, Verification of NBS Class 4 transport protocol, *IEEE Transactions on Communications*, Vol. COM-34, No 8, pp. 781–789, August 1986.

[3.13] S. Alfonzetti, S. Casale and S. Coco, Class 3 transport protocol specification, *Proceedings EUROCON 84*, pp. 131–135, 26–28 September 1984.

[3.14] AFNOR, *Interconnexion de systèmes ouverts. Service et protocole de transport en mode connexion*, AFNOR NF Z 70-002, Paris, September 1985.

[3.15] NBS, *Specification of a Transport Protocol for Computer Communications*, National Bureau of Standards, Washington, Vol. 1–6, June 1984.

[3.16] ECMA, *Standard ECMA-72. Transport Protocol*, ECMA, Geneva, September 1982.

[3.17] ECMA, *Standard ECMA-92. Connectionless Internetwork Protocol*, ECMA, Geneva, March 1984.

[3.18] ISO, *Open Systems Interconnection. Protocol for Providing the Connectionless-Mode Transport Service*, ISO, Geneva, Draft ISO/DIS 8602, 1985.

[3.19] P. Boudalier, INA 960. Logiciel de transport ISO pour l'architecture réseau local d'INTEL, *Electronique, techniques et industries*, No 20, pp. 33–44, June 1985.

[3.20] K.L. Mills, Testing OSI protocols: NBS advances the state of the art, *Data Communications*, pp. 277–285, March 1984.

[3.21] CCITT, *Data Communication Networks. Open Systems Interconnection (OSI). System Description Techniques. Recommendations X.200–X.250*, Red Book, Vol. VIII, Fascicle VIII.5, CCITT, Geneva, 1985.

INDEX

Index compiled by J.C.C. Nelson